THE BACKYARD CHICKEN BIBLE

the complete guide to raising chickens

ERIC LOFGREN

IR

Living Ready BOOKS
IOLA, WISCONSIN
www.LivingReadyOnline.com

Table of Contents

PREFACE 8

INTRODUCTION 10

CHAPTER ONE:
LET'S TALK CHICKENESE 12
    Learning a New Language 12
    Poultry Terminology 13
    Common Abbreviations 15
    USDA Poultry Classifications 17
        Cage Free 17
        Free Range 18
        Antibiotic Free 18
        Hormone Free 18
        Chemical Free 18
        Natural 18
        Naturally Raised 18
        Vegetarian Fed 19
        Certified Organic 19
        Certified Humane 19
        American Humane Certified 19
        United Egg Producers Certified 20
        Food Alliance Certified 20
        Animal Welfare Approved 20

CHAPTER TWO:
PREPARING TO KEEP CHICKENS 22
   Poultry Economics 22
   What Chickens Need 24
   What Chickens Eat 30
      Treats and Goodies 37
   Selecting a Breed 39
      Rhode Island Red 41
      New Hampshire Red 42
      Barred Plymouth Rock 42
      Delaware 43
      Ameraucana 44
      Leghorn 45
      Sex-Links 46
      Cornish Cross 46
   Building a Coop 48
   Nest Boxes 52
   Artificial Lighting 55
   Weather Extremes 59
   Avoiding Predators 66
   Gardening with Poultry 68
   Composting with Chickens 70

CHAPTER 3:
GETTING STARTED WITH CHICKS 73
    All About Chicks 73
    Setting Up Your Brooder 76
    The Life Stages of Chickens 78
        0 to 6 Weeks of Age 79
        6 to 14 Weeks of Age 82
        14 to 22 Weeks of Age 84
        Young Adults 86
    Broody Hens 88
        Surrogacy 92
        Breaking a Broody Hen 93
        Moving a Broody Hen 94
    All About Eggs 95
        How an Egg Is Formed 96
        Color Versus Taste 100
        Egg Freshness and Storage 103
    Incubating Eggs 105
        Some Basic Rules 106
        Egg Viability 107
        Heat Source 108
        Egg Turning 109
        Humidity 109
        Candling 110
        Lockdown 112
        Hatching 112
        After the Hatch 113
        Keeping Records 115
        Dos and Don'ts 115
        Uh Oh, What Happened? 116
        Fall Chicks 119

CHAPTER 4:
MAINTAINING YOUR FLOCK ... 123
- Hierarchy of Poultry Society ... 123
  - *Bringing in New Birds* ... 126
  - *Transitioning Chicks and Chickens* ... 130
  - *Changing a Flock's Dynamics* ... 132
  - *Poultry Personality* ... 133
  - *Aggressive Roosters* ... 136
- Bathing and Grooming Your Chickens ... 139
  - *Clipping Wings* ... 142
  - *Trimming Toenails* ... 145
  - *Trimming Beaks* ... 146
  - *Spurs* ... 148
- Molting ... 151
- Why Have They Stopped Laying? ... 153
- Genetics ... 155
  - *Sex-Links and Auto-Sexing* ... 158
  - *Crossbreeding* ... 160

CHAPTER 5:
POULTRY DISEASES AND AILMENTS ... 163
- Is There a Doctor in the House? ... 163
- Parasites ... 168
  - *Worms* ... 169
  - *Mites and Lice* ... 170
  - *Scaly Leg Mites* ... 172
  - *Stick-Tight Fleas* ... 174
- Ailments ... 176
  - *Curled Toes* ... 176
  - *Pasty Butt* ... 178
  - *Slipped Achilles Tendon* ... 180

  Spraddle Leg 180
  Vitamin and Mineral Deficiencies 182
  Flip-Over Syndrome 184
  Water Belly 186
Diseases 187
  Aflatoxicosis 187
  Avian Influenza 188
  Bumblefoot 190
  Chronic Respiratory Disease (CRD) 193
  Coccidiosis 194
  Fowl Pox 195
  Infectious Bronchitis 196
  Infectious Coryza 197
  Marek's Disease 198
  Newcastle Disease 200
  Respiratory Fungal Infections 201
  Sinusitis 203
Other Problems 205
  Botulism 205
  Ethylene Glycol Poisoning 206
  Impacted Crop and Flush 207
  White Birds Turning Yellow 210
  Cannibalism 210
Medications 212
  Vaccines 213
  Injectables and Ingestibles 213
  Tablets and Capsules 214
  Solubles 215
  Other Medicines 217
National Poultry Improvement Plan (NPIP) 218

APPENDIX A:
CHICKENS AND CHILDREN 223

APPENDIX B:
FUN WAYS TO USE YOUR EGGS 227
    Hard-Boiling Fresh Eggs 227
    Blowing Out Eggs 227
    Natural Egg Dyes 228
    Alternative Egg Dyes 230
    Egg in a Bottle 231

APPENDIX C:
ADDITIONAL RESOURCES 232

ABOUT THE AUTHOR 233

THANKS AND DEDICATION 233

INDEX 234

Preface

As a young boy, I had always dreamed of having chickens. I wanted a huge chicken farm with the big barn and hundreds of chickens wandering around at my feet. But, coming from a large family of the non-farming type, I was left to my own devices if I was to ever have chickens.

Our family never had much money when I was growing up so I couldn't just ask Mom and Dad for the funding for my crazy adventure. And with them not being of the farming mentality, I could not ask them for the advice I would need along the way. But what our family lacked in monetary standards was more than made up for in love and support in any adventure that we, as kids, were prone to follow.

My passion was for poultry and I didn't let anything stand in my way. My first chicken came from a neighbor who probably gave it to me so I would not pester him any more. That little hen was my pride and joy. As I got a bit older, I worked hard for the neighbor lady clearing her property of brush so that I could earn enough dollars to get my next set of chickens. I remember looking through all the catalogs that I could get to see all the wonderful products just for chickens. I wanted it all. I thought my chickens needed everything that those catalogs had to offer. I would make lists and more lists, and then I would work as hard as I could and then count my dollars only to find that I would come up short.

Chicks cost money and feed costs money. By the time I had bought the feed for the week, there was not much left of my dollars to buy the things on my list. I remember wanting a Marsh Turn-X Incubator. I don't even remember what hatchery it was that you could order them from, but that was always on my list. Silly as it may sound, the most important thing on my list was leg bands. I wanted my chickens to have those brightly colored numbered bands and they wouldn't be real chickens unless they had them, but they were always just out of reach. You had to order like fifty of them at a time and to me they were very expensive. I tried to make my own but that didn't work. I tried to work harder to earn extra dollars but it was never enough. My chickens would suffer because they couldn't have the colorful leg bands.

It wasn't until I was forty-six years old that my chickens finally got leg bands. Now my chickens no longer have to be just plain chickens; they are real chickens. But the funny thing is, though I now have enough bands to place one on every chicken if I so desire, out of all the chickens I have now, only one has a nice bright yellow numbered leg band. Still, I have carried that dream with me throughout life.

I have had to learn the ways of chickens the hard way, through trial and error. Believe me, there was plenty of error, but chickens are very resilient creatures and very forgiving. Every time I walk out the door they come running up to greet me. It might be because they think they are going to get a treat, but nonetheless they all come running. My chickens have provided me with endless hours of entertainment, but they have also given me heartbreak. I take it upon myself to learn their ways and their needs so that I might better serve them. Every day I continue to learn and every day they continue to teach me.

If your dream is to have chickens, then by all means follow that dream. And if your dream as a young child was to have brightly colored numbered leg bands on each of your chickens, then by all means come and see me and together we will make sure that dream comes true.

Introduction

The world of poultry keeping has come a long way since the days of Grandma having a few chickens down on the farm. Today's chickens are more than just birds that provide meat and eggs. They are the new pets. There are huge forum sites such as BackYardChickens.com dedicated solely to the raising of poultry. There are diapers you can get for your chickens so that they can be in the house or travel with you. There are even special treats and toys designed just for chickens. The days of a few chickens running loose on the old country farm are long gone. Now they are your typical house pets just as you would have a dog or cat.

Many cities, towns and municipalities are now rewriting their laws to allow chickens, specifically hens, to be kept inside city limits within small backyards. These are not little cities out in the middle of nowhere that are allowing this—no, these are major metropolitan areas such as Portland, Seattle, and Minneapolis. The poultry revolution is taking this country by storm and it is a wonderful thing.

Poultry have been domesticated for thousands of years and with literally hundreds of breeds and varieties to choose from, there seems to be a chicken to suit anyone's fancy. There are egg layers and meat producers, bantams and large fowl, feather footed and clean legged that all come in a wide variety of colors. There are the heritage breeds and the hybrids, the cross breeds and the mutts. It is all just a matter of what you are looking for in a chicken.

Chickens are not the dumb, nasty, foul-smelling creatures that they have been made out to be. Chickens are quite intelligent, considering their small brain size, and the only reason they would be nasty and foul smelling is if they were not properly cared for. Here at our farm, when people come to visit, they are amazed that there is very little, if any, smell and that there is a remarkable absence of flies. Granted, we are a small farm as we keep up to about 325 chickens, but we are large in the world of backyard poultry growers. We also hatch about 700 chicks each year in small incubators, which we sell to other local poultry people.

We try to take the logical approach to raising chickens. We get calls and emails from people all over the world wanting to know about all the different medications and all the specifics about raising and caring for chickens. It doesn't have to be as difficult as people want to make it out to be. Our philosophy is "Less is more"—less medicine, more healthy birds; less stress, happier chickens and happier you. People just starting out with chickens come into our little store and want us to set them up with everything they might need. They are shocked when I only hand them just a few items. They want to know about the rest of it and I tell them they don't need it unless something goes terribly wrong, and at that point I will get them what they need.

Raising poultry does not have to be a time-consuming, drive-you-crazy type of operation. It should be an enjoyable experience for all involved—for you, for your family and for the chickens. It is my hope that this guide will in some way help you to realize the enjoyment that so many have been able to find in their quest to have a small backyard flock. If you want two chickens or two hundred, the principles are the same. It all depends on you and the time that you have to devote to your birds.

1 LET'S TALK CHICKENESE

Learning a New Language

Humans have an incredible array of different languages and dialects throughout the world. Every species of animal also has its own form of communication and various dialects. And when these two paths cross, between the humans and the animals, new words, abbreviations, slang words and terminologies seem to emerge.

The first time you visit a poultry show you may feel you are in a foreign country with all the different languages going on around you. What does this person mean by a Blue Laced Red? Or that person over there talking about a BB Red OEGB? And this guy here, he's talking about how he got a Red Sex-Link from an SC RIR and an RC RIW. That doesn't sound too healthy to me.

As with any cooperation between the human world and the animal kingdom, there has developed a certain language unique to that species. With cattle, it is terms such as Brangus or Rocky Mountain Oysters. With goats, it is billy, disbudded and scours. Even with man's best friend there is AKC, CKC or whelping. So it would only be natural that there be a language specifically for chickens. It does seem, though, that in the world of chickens there are a lot more abbreviations used than in other animal species.

But what do they all mean? How do I tell what anyone is saying? The following pages will give you some insight into what I call "Talking Chickenese." By no means are these lists complete, but it is a good starter course for you with most of the common terminologies and abbreviations in use for everyday dealings with poultry people.

Poultry Terminology

Let's begin by explaining some common terms you'll come across in this book.

Chick - A newborn chicken up to six weeks of age.

Pullet - A female chicken under one year of age.

Hen - A female chicken over one year of age.

Cockerel - A male chicken under one year of age.

Cock - A male chicken over one year of age.

Rooster - A male chicken over one year of age.

Chook - English slang for a chicken.

Started pullet - A female chicken less than one year of age that has begun to lay eggs.

Laying hen - A female chicken over one year of age that is actively laying eggs.

Biddy - Term used for chicks and also for a hen.

Laying - The daily activity of laying an egg.

Sitting - The act of a broody hen while laying a clutch of eggs to hatch chicks.

Setting - The act of a broody hen after she has laid her clutch of eggs and begins the incubating and hatching process, a duration of approximately twenty-one days.

Broody - The act of a hen wanting to set on eggs.

Clutch - The number of eggs a hen sets on for hatching.

Brood - The number of chicks that hatch that a hen cares for and raises.

Bantam - A group of small breeds of chickens. Many are small versions of standard breeds. Bantams generally weigh less than three pounds.

Bantie or banty - Slang for Bantam but also used to distinguish small, mixed breed or mutt chickens.

Large fowl - The original standard size chickens as opposed to bantams. They are generally heavy-bodied birds that are four pounds or more in weight.

Breed - The specific breed of a chicken such as Plymouth Rock.

Variety - The varieties of a specific breed of chicken such as Barred, Blue, Buff, Black, White and Partridge, which are all varieties of the Plymouth Rock breed.

APA - American Poultry Association. The governing body that determines the standard of perfection for breeds of chickens within the United States.

ABA - American Bantam Association. The governing body that determines the standard of perfection for bantam breeds of chickens within the United States.

NPIP - National Poultry Improvement Plan. A Department of Agriculture voluntary program where they come out and test your birds for pullorum and typhoid diseases. Testing is a must if you are going to ship birds.

Capon - A male chicken (rooster) that has been surgically sterilized so that it cannot reproduce.

Purebred - The offspring of purebred parents that are of the same class, breed and variety.

Hybrid - A crossbreed, typically used for a purposeful crossing of two species to produce offspring with a unique set of characteristics.

Exhibition breed - Those breeds of chickens used primarily for show without regard to laying or meat qualities.

Utility breed - Specific breeds of chickens kept primarily for their egg laying and meat qualities.

Mixed breed - The result of breeding two different breeds of chickens. Generally referred to as barnyard mutts.

Sex-link - A genetic trait that creates a difference (usually in color) between males and females. Most often this is used to refer to traits that make chicks of different genders visibly distinct for ease of sexing. The term may apply to the gene or characteristic, or is often applied to hybrid crosses that display this characteristic, such as the Black Sex-Link.

Dual purpose - Those breeds of chickens specifically bred to provide both eggs and meat, such as the Delaware and Rhode Island Red.

Meat bird - Those breeds of chickens specifically bred for fast growth and broad chest designed primarily for eating, such as the Cornish Cross.

Point of lay - The age at which a pullet begins to lay eggs.

Molt - Part of a hen's natural reproductive cycle. After laying eggs for about a year, a hen loses her feathers and rests for a few weeks as new feathers grow in. This is called molting, or a molt, and it usually happens at the beginning of winter.

Common Abbreviations

There are many common abbreviations that are used in the world of poultry, especially if you are going to search on the forum sites. But you will also find that many people also speak in abbreviations. It would not be uncommon to hear someone say that they have "an SQ BBR OEGB that meets the ABA SOP," which translated means that they have a Show Quality, Black Breasted Red, Old English Game Bantam that meets the American Bantam Association Standard Of Perfection. In addition, there are some abbreviations that are used frequently online to describe human family members—they can be confusing if you're not familiar with them. Here are a few of the more commonly used abbreviations:

ABA - American Bantam Association
APA - American Poultry Association
BA or LORP - Black Australorp
BB - Buff Brahma
BBR - Black Breasted Red
BCM - Black Copper Marans
BF - boyfriend
BF - Bantam Fowl
BIL – brother-in-law
BJG - Black Jersey Giant
BLRW - Blue Laced Red Wyandotte
BO - Buff Orpington
BOB - Best of Breed
BOSS - black oil sunflower seed
BQ – breeder quality
BR - Barred Rock
BR - Bourbon Red (Turkey)
BSL - Black Sex-Link
BTB - Black Tailed Buff (usually Japanese bantams)
BTW - Black Tailed White (usually Japanese bantams)
CBOF - Cantankerous Bag of Feathers
CM - Cuckoo Marans
DB - Dark Brahma

DC - dear children
DD - dear daughter, dear dad
DE - diatomaceous earth
DF - dear father, dear fiancé
DH - dear (most times) husband
DM - dear mother, dear mom
DS - dear son
DSD - dear step-daughter
DSF - dear step-father
DSM - dear step-mother
DSS - dear step-son
DW - dear wife
EE - Easter Egger (Ameraucana Cross)
FIL - father-in-law
FWIW - For What It's Worth
GDW - Golden Duckwing
GF - girlfriend
GLW - Golden Laced Wyandotte
GPH - Gold Penciled Hamburg
GSL - Gold Sex-Link
JG - Jersey Giant
LB - Light Brahma
LF - large fowl
MIL - mother-in-law
MUTT – mixed-breed chicken
NPIP - National Poultry Improvement Program
NSQ – non-show quality
OEG - Old English Game (usually standards)
OEGB - Old English Game Bantam
POL - point of lay
PQ - pet quality
RIR - Rhode Island Red
RSL - Red Sex-Link
SCOVY - Muscovy Duck

SDW - Silver Duckwing
SIL - sister-in-law
SLW - Silver Laced Wyandotte
SO - significant other
SOP - Standard of Perfection
SQ - show quality
WCP - White Crested Polish
WTB - wanted to buy

Did you know?

The fear of chickens is known as alektorophobia.

USDA Poultry Classifications

The USDA uses certain classifications when describing poultry or eggs. These can become quite confusing and misleading. There are also other classifications that are given by the egg and poultry producers themselves as well as by animal welfare groups. Next time you go to the store, pay attention when passing the egg aisle and see all the different labeling that is now on the egg cartons just to try to gain your hard-earned dollars. But what do they all mean? Here is a rundown of the different classifications and terms that you may encounter.

CAGE FREE

This USDA classification pertains to egg producing hens. In a normal egg factory, hens are kept in battery cages where they are given a mere sixty-seven square inches of space to live out their lives. This area is smaller than a standard sheet of paper. Cage Free means that the hens are allowed to roam around an enclosed building and do not generally have any access to the outdoors but at least they can spread their wings and walk around. They can be force molted through starvation and the beaks can be clipped.

FREE RANGE

When you see this classification on egg cartons of processed poultry, you automatically think the chickens are spending their days wandering some beautiful farm without a care in the world. Truth is, all that this classification means is that the birds have some access to the outdoors. The USDA has no requirements to the amount of space or duration of time that these birds are allowed outdoor access. Producers are allowed to cut their beaks and force molt them.

ANTIBIOTIC FREE

Antibiotic Free is not an approved USDA classification. The USDA allows the terms "No Antibiotics Added," "No Added Antibiotics," "No Antibiotics Ever," and "Raised Without The Use Of Antibiotics." Using these labels, the grower must be able to provide documentation to prove that the birds were raised without antibiotics.

HORMONE FREE

The use of hormones is prohibited in the growth of poultry through federal regulation. Therefore, this label is not allowed to be placed on eggs or poultry unless the label also states "Federal regulations prohibit the use of hormones."

CHEMICAL FREE

The USDA does not allow this term to be used on poultry labeling. Federal regulations also do not allow the terms "Naturally Raised," "Naturally Grown," "Drug Free," "Residue Free," or "Residue Tested."

NATURAL

This USDA classification is generally only used on poultry and not on eggs. Growers are allowed to use this label if they can prove that the product does not contain an artificial flavor, color, preservative, or any other artificial or synthetic ingredient. But, the grower must also state why it is natural, such as no added color.

NATURALLY RAISED

You can almost envision a large flock of beautiful birds wandering carefree around some well-manicured Vermont farm, spending the day doing what chickens do best. Back to reality folks: This term is not a USDA classification. It is a voluntary program that is used to show that the birds were raised without growth proponents and antibiotics and that they are not fed animal byproducts. It has nothing to do with how they are kept.

VEGETARIAN FED

The USDA does not regulate this term but it brings to mind some lavish farm where the chickens are spoiled with lush greens and vegetables. Really, all it means is that the chickens were not fed any animal byproducts or dairy products, but the growers must have proof to support the claim.

CERTIFIED ORGANIC

This USDA certification is one of the hardest to obtain due to the vast amounts of paperwork involved. The birds are kept un-caged and fed a strict diet of organic, vegetable-based feed free from pesticides and antibiotics. Each ingredient in the feed must have its own certification as to being organically grown. Debeaking and force molting are allowed and it is unclear if access to the outdoors, open pasture, and natural light are required.

CERTIFIED HUMANE

The USDA does not regulate this term. It is a voluntary program administered by Humane Farm Animal Care. The birds are allowed to roam freely inside an enclosed facility but are not required to have access to the outdoors. The birds must be allowed to display natural behaviors such as nesting, perching and dust bathing. There are certain regulations that have to be met, such as stocking densities, perch area and number of nest boxes available.

AMERICAN HUMANE CERTIFIED

This program is available through the American Humane Association and allows for both caged and cage-free keeping of birds. Birds under this program are kept in very much the same manner as standard commercial battery cage houses, but the cages must be slightly bigger, giving each hen space about the size of a legal sheet of paper. Birds cannot be force molted by starvation but their beaks can be clipped.

UNITED EGG PRODUCERS CERTIFIED

This term really means nothing as the hens are kept under typical battery cage systems. It only means that the hens are not force molted by way of starvation.

FOOD ALLIANCE CERTIFIED

This certification is pretty much the same as the "Free Range" classification of the USDA. Birds must be allowed to roam freely with access to the outdoors and natural light. There are specific requirements for stocking densities, perches and nest boxes. Forced molting is not allowed. This program is administered through the Food Alliance.

ANIMAL WELFARE APPROVED

This program has the highest standards of any of the poultry programs. This program allows the birds to be kept in the most humane of conditions with being cage free and having continuous outdoor perching access. There are strict regulations for nest box availability, perches and stocking densities. The birds must be allowed to molt naturally and their beaks cannot be clipped. This program is sponsored by the Animal Welfare Institute.

You know you're addicted to chickens when . . . you stop mowing the lawn so you can catch the bugs in the grass for the chickens.

External parts of a chicken

2 PREPARING TO KEEP CHICKENS

 ## Poultry Economics

I have to stop myself from laughing every time someone mentions that they are going to raise a few hens or some meat birds to provide extra income for their family by selling eggs to their neighbors and an occasional fryer for a Sunday dinner. You have to raise chickens for a reason other than to get rich. Yes, it is true that you can supplement your feed bill by selling eggs and meat, but that is about as far as it goes. The rare instance of this would be if you are going to raise show-quality birds to sell to others, but even then you would be hard pressed to make a living at it.

Let's look at a typical situation. You want to get some chickens so you scour the poultry catalogs looking for what breed best suits your desire. Before the chicks arrive, you must provide them with a brooder area, feed, water and a heat source. You run out and buy a large plastic tote, feeder, waterer, heat lamp bulb and lamp shroud; total cost approximately $35. Chicks consume large amounts of feed as they grow, so you buy a fifty-pound bag of chick starter; there is another $15. You get your brooder all set up and order your chicks. Most hatcheries require that you buy a minimum of twenty-five chicks. With shipping, they work out to about $3.50 each, so tack on another $87.50. The chicks will soon outgrow the plastic tote, so you will need to build them an adequate grow-out pen; with minimal lumber and chicken wire, that adds another $75. Once your birds reach about sixteen weeks, it will be time to move them to their permanent coop. There will be larger feeders, waterers, nest boxes, roosts, coop and pen. For economy's sake, let's just say you expand their grow-out pen, so add a mere $150. With the cost of feed today, to get a hen to laying age of twenty-two weeks, the average feed cost is approximately $18.50 per hen. That would amount to $462.50 in feed. That brings you up to $825 in expenses before you get your first egg. Yes, there are cheaper ways of doing things to help bring down the cost a bit, such as recycled materials and free ranging your birds as much as possible, but they do not cut it significantly.

After a couple of months of laying pullet-sized eggs, your hens are laying well and you have an excess of eggs. With a great laying hen, it takes about four pounds of feed for her to lay a dozen eggs. At the going rate for a fifty-pound bag of layer feed, that will cost you about $1.20 for her to produce that dozen eggs. Now, most states require you to use new egg cartons to legally sell eggs, so add in another 32 cents for each carton. Depending on your location, farm fresh eggs generally sell for about $2 a dozen. That leaves you with a profit of 48 cents for each dozen eggs. And this does not include any medications that you might have to provide to your birds during the rearing stage.

A good laying hen can lay a dozen eggs in about sixteen days. Twenty-five hens at twenty-five dozen eggs at 48 cents a dozen profit will net you about $12 profit every sixteen days. You are thinking, wow, in a year's time that would be an extra $273.75 in my pocket. At that rate it will only take about three years to pay back the initial investment of getting your girls to laying age, and by that time, your hens will be old and need to be replaced. There is one slight problem in all of this, though: hens do not lay all year long unless you have them under artificial light and force molt them.

Chicks bought in the spring will be ready to lay about August. By the time they are laying nice large eggs, fall is here and then winter when the hens will all but stop laying. Then you have to wait for the following spring before they start laying again full time. That is more feed that you have to figure into the equation. Springtime rolls around and now five of those heritage breed chickens you have so dutifully raised have decided to go broody and are now setting on eggs instead of producing eggs. Arrggg! Where does the madness stop? So, as you can see, you are not going to make a living by selling eggs.

What about meat birds? Today's hybrid meat birds grow fast. Extremely fast in some instances. It is not uncommon to get an eight-pound bird in less than fourteen weeks. The commercial growers are getting nine-pound birds in eight weeks, but meat birds consume huge amounts of high protein feed to compensate for their rapid growth. At current feed prices, it costs approximately $22 to get a bird to market weight. If you were able to raise thousands of birds at a time and buy your feed in bulk, then the price per bird would drop considerably. Then you have to find a market for your birds. Chicken in the store is cheap, so it is very difficult to find someone willing to pay you $22 for a bird, let alone trying to make a profit off of them.

So, how does a person actually make any money at raising chickens? You really don't, but you can come pretty close by using a variety of plans. Newborn chicks can be a moneymaker if you have an outlet for them. Chicks generally sell for around $2 each if they are purebreed. Some of the fancier breeds sell for more. With a small tabletop incubator and a standard hatch of about thirty chicks, you could see about $60 profit every twenty-one days. If you were hatching good quality, potential show chicks, then that could easily jump to $150 or more. You could even just sell the hatching eggs. Think of it this way, you could sell a dozen eating eggs for $2 or, with a good rooster and nice hens of a desirable breed, those eggs could now sell for $10 to $12 a dozen. It is all a matter of what market opportunities you have available in your area. Think outside the box of just selling eggs and meat. Expand your horizons and you may be able to make the birds at least break even.

You have to raise chickens because you want to, not because you want to make a living at it. By raising your own chickens, you get the benefit of knowing what goes into them as well as what comes out of them. You have to do it for the enjoyment that chickens can provide. Just know that with raising chickens you get the added benefit of farm fresh eggs and meat.

What Chickens Need

You have read all the books, you have looked at all the fancy chicken catalogs, you have talked with other folks who are raising chickens, and so now you have decided to take that fateful step into backyard chicken raising. But there is still that lingering question in the back of your mind: what do I really need to do on my part to raise chickens? After all, everyone out there has a different opinion.

Many times throughout the year, especially in the spring, I get phone calls and e-mails from folks all over wanting to know what they need to do before they get their chickens. Most of them have these grandiose ideas of what they think they need for their chickens, and that is perfectly fine. What I tell them is don't over-think the process. Chickens should be fun and easy to take care of or the novelty of the whole thing will wear off really quickly. These folks that build or buy the really fancy coops—go look at them in six months and see what that coop looks like then and ask them how many hours a day they spend trying to keep that fancy coop clean. Or the ones who insist they need every medication—go ask them how many of those medications they throw out because they expire before they ever use them.

Granted, you should be prepared before you bring any chickens home, but really it can be much easier than most people make it out to be. If it is not fun then you will never understand the enjoyment that chickens can bring to your life.

The two main aspects that need to be considered are if you are going to start with day-old chicks or grown birds. Let's look at grown birds first, that is birds over six weeks of age, seeing as how even if you get day-old chicks, they will soon become grown birds.

Looking at the very simplistic side, you will need a cage, a coop with a nest box and roost, a feeder and feed, and a waterer and water. These should all be in place before you ever bring a chicken home. But is it really that simple? It can be if you want it to be.

Space is your main consideration. How much space you have to devote to raising chickens will determine how many chickens you can have, or how many chickens you want will determine how much space you will need. Just for simplicity's sake, let's say you are planning on getting heavy breed layers such as Barred Rocks or Rhode Island Reds. A heavy breed bird such as these requires one foot of roosting space, four square feet of coop space, and ten square feet of run space per bird. You live in an urban neighborhood and want to set up a coop and pen to one side of your property. You have a small existing shed that is 4' × 6'. The shed would limit you to six birds, and so your pen area would need to be sixty square feet. Always keep in mind when planning a coop and pen that the more area the birds have, the less problems you will have with them and the healthier they will be. Inside the coop you will need at least one roost that is six feet long. In addition to the coop and pen, you will need one feeder, one waterer, a couple of nest boxes—and you will be good to go.

Now let's expand on this idea just a bit. The coop you build, buy or repurpose needs to be suited for your local weather. You would not want to necessarily use an all-metal shed as your coop in the hot south because during the summer months it will be like an oven inside and at the least you would severely stress your birds from the heat. Likewise you would not want a simple metal shed in the northern areas because during the winter it would be pretty much like an icebox.

Your coop should have good ventilation but not subject the birds to drafts. It should also be able to protect the birds from predators during the night while they are sleeping and it needs to be properly insulated to handle cold, or ventilated to handle heat. Your coop will also need a door to the outside so that you may clean it on a regular basis and also for gathering eggs if your nest boxes are inside. It should also have a smaller door opening for the birds to pass in and out of from the coop to the pen. This door should be able to be closed off at night for protection from predators. This is discussed in greater depth in the Building a Coop section later in this chapter.

The pen, or run, should provide plenty of space for the birds to easily move around and not be crowded. Your pen should be built to adequately protect your birds from predators both on the ground and in the air. The pen can be built in various ways and with a wide variety of materials. This will depend on what you have available to you. At least part of your pen area should be covered to allow the birds an area to get out of the sun or rain. Part should be left uncovered so that the birds can have necessary sunbathing time. An easy design for pen construction consists of a simple wooden frame with 2" × 4" welded wire mesh on all sides and also extends out a minimum of one foot around the perimeter of the base to keep digging animals from trying to get in under the wire. The side wire is then covered up a minimum of two feet from the base with a smaller opening wire such as ½" × ½" hardware cloth the keep out most ground base predators. Over the top of the pen use wire mesh or at a minimum bird netting to help keep out airborne predators. The pen should be firmly attached to the coop.

The size of your coop will help determine what other amenities you can have for your birds. Inside the coop will need to be a roost, which can be made from just about anything. A 2 × 4 on edge with the corners rounded makes a nice sturdy roost that is easy for the birds to hold on to. The roost should be mounted no higher than three feet off the floor for heavy-bodied birds. This is low enough for them to easily fly up onto at night and not too high to cause leg and foot problems when they jump down in the morning.

If your coop is large enough, you will be able to place the feed and water containers inside. If not, then you will need to place them out in the run. Make sure that both are placed under cover to help keep the feed dry, and also so the birds do not have to go out in the weather to eat and drink. It is best if you are able to hang both the feeder and waterer to help keep them clean and to not waste feed. Hang them about chest high of the birds. They will be able to reach the feed and water but not able to scratch it out and be less likely to soil it. If you can't hang them, then place a block under each to raise them off the floor.

The age of your birds will determine the correct feed to use. For birds under eighteen weeks of age, a starter/grower feed should be used. For birds over eighteen weeks of age, a layer feed, either pelleted, crumble or mash, can be given for laying birds, or a flock raiser or higher protein feed can be given to meat-type birds. The amount of feed that is consumed will be determined by your choice of care methods. Some folks only feed their birds in the morning and at night. I choose to feed mine free choice all day long. This will depend a lot on whether your birds have access to free range or if they are constantly cooped, and also on the desired growth rate of your birds.

Water is of the utmost importance for your birds. Your birds should never run out of fresh cool water at any point in the day. Your weather conditions will help determine the amount of water they drink in a day. This will need to be closely monitored and additional waterers added if needed. The water should be changed at least daily to keep it fresh, and more often in hot climates.

You will also need nest boxes for the girls to lay their eggs in. The nest boxes should be in the coop or mounted to the outside of the coop with holes cut through for access. If mounting the boxes inside, they can be something as simple as a milk crate filled with some straw, or you can build boxes. The box should be no smaller than 12" × 12" preferably 14" × 14" for heavy-bodied girls. It would be advisable to create some sort of sloped roof over the top of the nest boxes to discourage roosting in or on the boxes. More about nest boxes is covered in the Nest Boxes section found later in this chapter.

The one other item that may need to be included in your coop/pen set up, depending on your soil type, is a dust bath box. Birds of all natures take baths. This is to help clean the oils and dirt from their feathers and to also help rid their bodies of menacing parasites. They will then groom themselves to straighten their feathers and to also spread new oils secreted from a gland near their tail. Chickens take their baths in the dirt by scratching up some loose dirt, then lying down in it and with varied body motions, fluffing it through their feathers. This is a natural process for them. Depending on weather conditions, they may take a bath every day. If your ground is hard or heavily covered in vegetation, your birds will need a supplementary dust bathing area. Many folks will use a large, short-sided wooden box filled with fine sand as a dust bath for their birds. Some folks may even add some diatomaceous earth to the sand to provide better parasite control. Dust bathing is a communal activity, so when considering the size of box, be aware that more than one bird may be bathing at a time.

Any of these processes may be expanded on to your individual liking and determined by your local climate conditions and the number of birds you plan to keep. Now, let's add in the possibility of starting with day-old chicks. The process is nearly the same but with a few added factors.

Whether you order chicks from a hatchery or get them from the farmer down the street will not matter; what matters is how you care for them once they are home. Your new chicks will need to be in a brooder. A brooder can be as simple as a cardboard box or as complex as you want to make it. I use the largest clear plastic tote I can get, which is the 105-quart size. This works well for about twenty-five chicks for the first two weeks before I move the chicks to an outside brooder area. Whatever you choose to use as your brooder, it needs to be draft-free, well-ventilated, and able to retain up to 95ºF (35ºC) of heat. Indoors, I use the plastic tote; outdoors, I use a large wooden box that I built out of plywood.

You will need a heat source. The most common source of heat is a heat lamp suspended over the brooder area. What you are aiming for is to have a constant temperature of 95ºF (35ºC) at about two inches off the bottom of the brooder. Depending on your conditions, this may be acquired with the use of just a regular light bulb or you may need one of the heat lamp bulbs available in either 125-watt or 250-watt sizes. I find for my brooder that a common 85-watt floodlight works really well; it throws plenty of heat but does not use nearly as much electricity as the larger bulbs. One thing to consider is that the larger 250-watt heat lamp bulbs throw massive amounts of heat and there is the inherent danger of baking the chicks, or worse yet, setting something on fire. I don't know how many times I have heard of someone burning down their coop or barn because they were using a heat-lamp bulb. If you need to use a 250-watt heat lamp to keep the brooder at 95ºF (35ºC), maybe you should look at a different style of brooder.

Your brooder should also have some form of protective lid made of wire mesh to help keep out unwanted pests and curious pets.

On the floor of the brooder, you will need some sort of absorbent material. For day-old chicks, it is best to place several layers of paper towels for the chicks to walk on at least for the first four to five days. This will provide them with a non-slippery surface so that they do not end up with leg or foot problems. It allows the chicks to find feed easier and also keeps them from digesting other types of materials until they learn where the food dish is. After four or five days, you can switch to a layer of pine shavings or other suitable material. Never use cedar shavings with your birds, as the oils in cedar are potentially harmful to chickens.

You will also need a small feeder and waterer. In the plastic tote brooder, I can easily place a one-quart feeder and a one-quart waterer. These are easy for the chicks to learn to eat and drink from. The number of chicks you have will determine the number of feeders and waterers needed. Feeders come in various sizes and configurations. Some are round, while others are long trough-like fixtures. It doesn't matter what you use as long as there is adequate space for the chicks to feed. The waterer needs to be small enough that the chicks do not fall in or get stuck and drown, but large enough to provide plenty of fresh water. You may find that you will be changing the water two or three times a day just to keep it clean and fresh. As for feed, you will want to provide the chicks with a supply of good quality chick starter feed.

You won't need a thermometer except for maybe at the initial set-up. The chicks will tell you if the temperature is correct. If they are all huddled under the light, then they are too cold. The light will need to be lowered or a larger wattage bulb used. If the chicks are avoiding the light and are mushed out to the sides, then the light is too hot and needs to be raised or a lower wattage bulb used. The chicks should be flitting about and resting all over the brooder floor. This will tell you the temperature is just right. The heat in the brooder will need to be lowered by 5 degrees for each week the chicks age until ambient air temperature is reached and the lights can then be turned off. This is discussed in further detail in chapter three.

Chicks grow fast, so you will need to have an expandable brooder or multiple brooding areas. As mentioned before, I start my chicks in a large plastic tote. At about two weeks of age, or earlier if the weather is cooperating, the chicks are moved to an outdoor brooder box that is 3' × 4' with eighteen-inch high sides that has a good layer of pine shavings on the floor. This is large enough that I can place two lamps over it if I am brooding during the winter, and also gives plenty of room for the growing chicks. At this time they also get a larger feeder and multiple waterers. The chicks will remain in this box for up to two weeks, depending on the weather conditions, and then they are moved to a pen, in which I have a lamp to help with keeping them warm mainly at night. By the time the chicks are six weeks of age, they should no longer need an additional heat source unless the weather is cold. I force my chicks along, but I also live in the warmer south, where they rarely ever need additional heat past five weeks of age. During the hot summer months, I can get away from heat at about four weeks of age. Once the chicks are out from under heat, they can be placed in the pen for continued grow out.

So whichever is your choice, whether to start with grown birds or with day-old chicks, make sure that all your equipment is in place and ready prior to the arrival of the birds. This will minimize the stress on the birds as well as on you. Again, do not over-think the process. Chickens are quite the minimalists and don't require a bunch of fancy things. The simpler you keep it, the easier your life will be, and the more enjoyment you will receive from raising chickens.

Did you know?
Chickens' taste buds are not well developed. They can distinguish between different tastes but are not affected by spicy foods like we are. Therefore, it is possible to mix such spices as cayenne pepper into their feed to aid in curing intestinal problems.

What Chickens Eat

What you feed your chickens is just as important to their well-being as any other part of raising chickens. There are many schools of thought as to what is the proper diet for the backyard flock. In the commercial industry, feeds are micromanaged to give the absolute best results for body weight gain or maximum egg laying, and these feed mixture formulas are some of the most closely-guarded secrets of the modern world. For those of us who raise backyard flocks, we can just run to the local feed store and pick up a bag of feed, but it does entail a little bit more thinking than that.

There are a few things to first consider, such as how are you going to raise these birds—cooped, free range, natural, organic or just as a flock of backyard chickens? There are different feed mixtures and textures, and each feed mill seems to have its own formula—so where do you start? Well, let's start at the beginning with chicks.

In commercially-prepared feeds, you have a couple of options available for new chicks. The most common is medicated and un-medicated chick starter/grower. Both of these come in a fine crumble that is easier for a chick to eat and digest. The medicated feed has a coccidiostat such as Amprolium added to the feed to help combat coccidiosis in young chicks. If you are of the thought that you want to raise your chicks as naturally as possible, then you might want to go with the un-medicated variety, which does not contain a coccidiostat. Each of these feeds works well for chicks from the time of hatch up to about eighteen weeks of age.

Next in the feed line is what is known as flock grower. This feed has a little bit different composition from the chick starter/grower with generally a bit more protein and usually does not have a coccidiostat added to the feed mix. This feed can be used for chicks that are over eight weeks of age and up to point of lay, which is about eighteen weeks.

Once pullets reach sexual maturity—anywhere between seventeen and thirty-two weeks, depending on the breed—they should be switched to a layer feed mix. This can be determined by a pullet squatting for you, a rooster starting to try to breed the pullet, or the appearance of the first egg.

Commercial layer feeds are carefully developed to be a complete feed for the pullets and hens to be used throughout their laying years. They come in either pelleted or crumble varieties. Protein levels are at 16 percent or higher and the feed has added calcium to aid in shell production. Layer feeds do not come in medicated varieties because of consumption of the eggs.

If you are choosing to raise meat birds, such as the Cornish Cross, there are commercially-prepared feeds specifically designed for their fast growth rates. These feeds will have high protein levels, usually in the 24 to 28 percent range.

Game Bird feed is just that, specially formulated for high-energy game birds such as quail, chukar and pheasants. There is a big misconception that this feed is formulated for chicken breeds such as Old English Game and Modern Game. Game Bird feed is formulated with around 30 percent protein for fast growth and feathering in sporting-type breeds and is not recommended for use with chickens.

There is also a feed known as All Flock. This is a general-purpose feed that is formulated to be used with a variety of poultry, such as turkeys and ducks as well as chickens. This is a great option if you have more than one variety of poultry and you don't want to buy a variety of feeds, or if the poultry are all free-ranged together. You will not get optimum results from your laying hens or your meat birds, but this lack of performance may be offset by the ease of using only one feed.

If you are fortunate enough to have a feed mill near you, then it is possible to have feed custom mixed to your specifications. They have their normal formulas for feed but will usually mix to your specifications if you purchase a minimum of one thousand to two thousand pounds of feed at a time. Some of the feed mills only produce a flake type feed, while others have the ability to produce a compressed feed in either pellet or crumble form. Being able to work with a feed mill can give you certain advantages when custom mixing your feeds. For those folks that are trying to raise organic birds or as naturally as possible, custom mixing gives you the opportunity to know just what is going into your birds.

Some folks develop their own custom mix of feed at home. This is OK if you have a working knowledge of the specific vitamin, mineral, and nutrient needs of chickens and can closely monitor their health and growth rates. What happens many times for folks who mix their own feeds is that their birds will become deficient in one or more essential vitamins or minerals and their birds will suffer from various ailments due to these deficiencies.

There are a multitude of different feed manufacturers out there and each one of them has their own formula for creating what they claim is the perfect feed. Years of research have gone into each formulation and some are better than others. It is a matter of finding which feed works best for you. Don't be afraid of changing feed brands until you find the one that your birds thrive on the best.

Around the farm here, I use commercially-prepared feeds by a major feed producer and have great results. I start my chicks on a medicated chick starter/ grower and then switch to a layer feed. The only variation is that my bantam breeds get crumble and my large fowl get pellet varieties once they reach maturity. Well, there are other variations to my feeding program and I will go into those later.

I don't know how many times I have been contacted by folks who are complaining that their birds are not producing eggs like they should and are just not thriving like other people's birds are. After going through a process of elimination as to what might be wrong with their birds, it is discovered that they are only feeding their birds scratch feed.

Scratch feed is simply a mix of three or four different grains, usually corn, milo, soybean, oats or various other grains. Scratch feed is not a complete balanced diet for chickens and should not be fed as such. Scratch feed should only be fed as a supplement to their regular feed and be given on a limited basis as a treat. Scratch feed should also not be fed to chicks under eight weeks of age.

A legitimate question arises quite often when it comes to feed and mixed-age flocks. What feed do you use if you have young birds out with older birds because you can't keep them all separated during feeding? And what do you do if you have a hen with a clutch of new chicks running around the barnyard mixed with all the older birds? There are lots of schools of thought on this, and each person finds what works best for them and their birds. There are the purists out there that have every age of bird kept separated until the point of maturity when they can be integrated, and each age bird gets its own feed. But this can take lots of different pens and lots of extra management within your flock. Then, on the other extreme, there are those that are of the thinking that if they put whatever feed out there, the birds will eat it if they get hungry enough.

Like I have said, I do things a little bit more unorthodox than some may feel comfortable doing, but it is what has worked well for me over the years. My feed program goes like this: All of my artificially-hatched chicks, meaning incubated, start out on a medicated chick starter/grower. They remain on this ration until they are about eight weeks of age when they are integrated into the main flock. They are still allowed to roost in their own pen at night, so I keep a feeder full of medicated chick starter/grower in their pen that they can eat from in the mornings before they go out with the rest of the flock and also eat from at night when they go back to their pen to roost. The rest of the day, when they are out with the bigger birds, if they get hungry they eat layer feed like the rest of the birds.

The chicks that are naturally hatched, meaning hatched by a hen in breeding pens, get a 50/50 mix of chick starter and layer crumble. I have found that this suits the chicks as well as the older birds, providing them with the necessary nutrients that they both need.

The chicks that are hatched naturally by a hen out with all the other free-range birds get to eat what all the other birds get and what the mama hen finds for them.

All three of these methods have worked very well and I do not have growth or nutritional issues with any of them. This may lead one to believe that it does not matter what you feed to the birds, but that is not correct thinking. With a naturally-hatched chick, the mama hen teaches the chick what is good to eat to get a complete diet that works well for growth. In an artificially-hatched chick, it is up to us to provide the chick with a balanced diet, and that is where commercially-prepared feeds come in.

The younger birds get fully integrated into the flock at about twelve weeks of age. At this time they are eating just layer feed and will remain on this for the rest of their lives. Other than commercially-prepared feed, my birds get the occasional scoop of scratch feed as a treat. This I only give to them once or twice a week, depending on how generous I feel. My free-range birds also get all the table scraps from home meals and my cooped breeder birds get the occasional treat of either fresh popped popcorn or bread. My girls stay fat, happy and sassy. They will gladly tell you how I starve them and never give them treats, but that is just not the case and I have the feed bill to prove it.

Another question that arises is how much to feed a chicken. There are also different schools of thought on this. My birds all have feed in front of them twenty-four hours a day, seven days a week. Some people will only give their birds enough feed in the morning to where they empty the feeder by the time they are done with the morning meal and do the same at night. The rest of the day the birds are on their own to find what they can. This works fine for free-range birds where they can get out and naturally forage, but for caged or cooped birds, this can put undue stress on them, causing them to not perform as they should and to be more prone to disease. How much feed a chicken eats in one day is almost impossible to tell. This all depends on the breed of chicken, age, and environmental factors. As a general rule of thumb, a standard laying hen will eat approximately $\frac{1}{3}$ to $\frac{1}{2}$ cup of feed a day; bantam breeds will eat less and meat birds will eat more.

Chickens eat the most just after waking in the morning and just before they go to roost at night. They gorge themselves on feed before roosting to provide their bodies with a source of nutrition through the night. By morning their crops are empty and they gorge themselves once more to get their bodies going again. Feed gives a chicken energy and also produces heat in their bodies. Chickens will eat much more during the winter than they do during the summer. Many people will

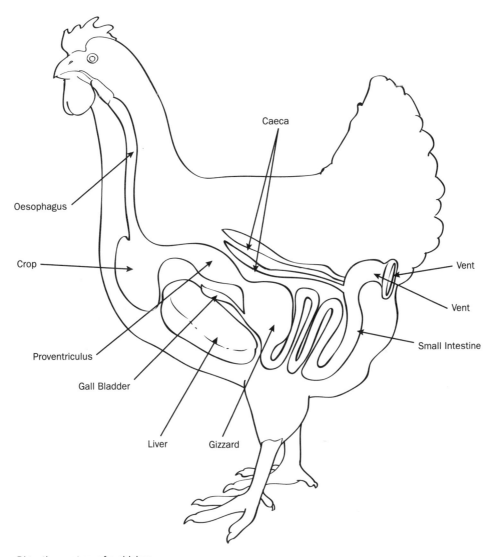

Digestive system of a chicken

adjust the protein levels in the feed during the summer and winter to help the birds cope better with the hot and cold temperatures. During the summer months, I feed my birds a standard 16 percent protein mix feed. During the winter months I boost the protein levels to around 18 to 20 percent, depending on the weather. If we are going to have an exceptionally cold night, then many times I will supplement their evening feed with a kitten feed that is 30 percent protein, which helps their body produce more heat and, therefore, helps to keep the bird warmer as it sleeps.

This brings up a simple observation. If you are having problems with your birds overheating during the summer months and you have provided them with plenty of shade and cool, fresh water but they are still overheating, try changing their feed to a lower protein level. Likewise, if you are having birds that are struggling during the winter to stay warm, try adding protein to their diet or change to a higher protein feed around 20 to 24 percent.

I know, you are still thinking about the paragraph above where I mentioned kitten feed. These are chickens, not kittens. Why the heck would you give a chicken cat food? There are times when your birds will need a temporary increase in protein levels such as during molt, cold temperatures, or illness. Instead of changing their normal diet, you can simply supplement their feed with something with high protein. There are many things out there that will do this—black oil sunflower seeds, beans, tuna or other fish and so on—but I like to use kitten food the best. A couple of feed manufacturers make a really small kitten kibble that is nearly the same size as pelleted chicken feed. Protein levels run from 28 percent to 40 percent and the chickens seem to be able to digest it well. If I know that we are going to have one of those cold nights, I will put out a small bowl of kitten kibble for each breeder pen and also for the free-range birds. A breeding pen of ten birds will get about a cup of kibble. This helps their bodies produce some extra heat to get them through a cold night. I also give the birds kitten kibble if they are going through a hard molt. I give them the same amount, one cup per ten birds, about three times a week. This added protein helps them to grow in their feathers quicker.

One concern that comes up from time to time is a bird getting too much protein in its diet. This can cause permanent damage to the bird. If you are going to be feeding a high-protein diet to your birds, watch for them to start walking on their tiptoes or walking with stiff legs, kind of like a high-step march. This is an early indicator of too high of levels of protein. If this is the case and your birds are showing these symptoms, back down on the available protein and the symptoms generally subside within a couple of days.

On a separate but worthy note, never store more feed than you can use in about two weeks' time. I know it is tempting when you get coupons for dollars off of a bag of feed to run out and buy hundreds of pounds of feed for just a small flock of birds. What happens is that the vitamins and minerals in the feed start to lose their potency after about two weeks especially vitamins A and E. Deficiencies of these two key vitamins can cause all sorts of issues with your birds. Depending on how your feed is stored, you can also end up with mold in your feed, which can cause af latoxicosis (food poisoning) in your birds. Stored feed can also become riddled with weevils. Feed is a grain-based product, so the potential of bugs in the feed is pretty high. These issues are also a good reason to buy your feed from a reputable dealer who sells a volume of feed so that you know you are always getting the best quality and freshest feed.

With any feeding program, whether it is commercially prepared, feed mill or a home-grown recipe, find out what works best for your birds making sure to provide the correct balance of vitamins, minerals and nutrients to keep the birds healthy and happy.

You know you're addicted to chickens when . . . you spend more time preparing meals for the chickens than you do for your own family.

TREATS AND GOODIES

People often ask what types of treats they can feed to their chickens. There is not much that a chicken won't eat. They truly are pigs with feathers. But there are a few things you have to be aware of when feeding your flock treats.

Small chicks should not be given treats. Their bodies are growing so fast that they need to have proper nutrition to maintain proper growth. Once your birds reach about twelve weeks of age, then you can start introducing them to treats if you wish. All treats should be given in moderation.

Chickens can pretty much eat anything that you do. Any leftovers from the refrigerator can go to the chickens as long as there is NO mold. You will not want to feed your birds such things as peelings, especially from uncooked potatoes. Most peelings they will not eat anyway. Last night's spaghetti? They love it. Salads, fruits and vegetables? Toss it out there. Leftover bread or popcorn from movie night? You better watch your fingers.

What we find is that the younger birds don't really know what to do when we toss them different treats, but the older birds have it all figured out. If we come out the door with any type of bucket or pan, they come running.

But what about dishes that contain meat? Or the leftover carcass of a turkey? Many people do not want to give their chickens any type of meat, but chickens actually love it. Just make sure that it is cooked first. Chickens are direct descendants of Tyrannosaurus rex, who was in fact a meat eater. We feel that the added animal protein does them good. And no, it won't cause your chickens to become cannibalistic. Actually, it is potentially more harmful to give your flock worms that you just dug up in the garden than it is to give them a leftover hamburger from yesterday's barbecue. Worms and bugs pick up diseases out of the soil that can be passed to your birds, whereas a hamburger has been cooked to kill off any potential diseases.

As with anything, treats have to be given in moderation. If you give your chickens too many treats, they will not want to eat their regular feed; egg production will drop, and they might not be getting the proper nutrition that they need. So limit the amount and types of treats that you give them. Scratch grains, for instance, are a great treat for chickens but there is very little nutritional value there. It is not something that you would want to feed too much of and you certainly do not want to make it their sole ration.

There are commercially-prepared treats for chickens. There are various seed blocks as well as dried mealworms. These are fine to use as feed supplements. And any treat should be just that, a supplement.

Fall is our chickens' favorite time of year. With the leaves falling to the ground and all the pruning of the bushes, all of this goes into the main chicken pen to be picked over before it goes to the compost pile. This is a huge treat for them. We do not worry too much about what goes into this pile except for long grasses. The long grasses can cause the birds to get impacted crop. But the birds will pick through the pile looking for the tastiest morsels. They will leave alone the plants that do not taste good to them. Yes, there are a few plants in different areas of the world that are harmful to poultry and you need to check with your local county extension office to see what those might be.

One of the more common treats that we feed our chickens, besides all the leftovers, is popcorn. We buy fifty-pound sacks and pop enough for each pen to have a couple of cups full. Noodles, the chickens go crazy for them. For a good laugh, cook up a pound of spaghetti (unbroken) and put it out there for the chickens. They grab a noodle and go running. All you see is long noodles flapping in the breeze. We get stacks of bread from the local bakery store and save it in the freezer. Take out a couple of frozen loaves and chop it into little pieces and throw it out like scratch.

But our birds also get wise to us. They know if they complain loud enough that we will cave in and give them treats. But this year I am one up on them. Most of our birds eat pelletized feed, and they enjoy it if there is a bit of crumble left over from feeding the bantams, which we throw to the other birds. So this year, I have taken a large feeder and I put it out with crumble feed in it. They think they are getting a real treat but really they are getting the same layer ration only in a different form. I am just wondering how long it will take them to catch on.

Treats are great for your birds, but just remember, absolutely no moldy foods, as it will potentially cause botulism. And, of course, give them treats in moderation.

You know you're addicted to chickens when . . . you start to grocery shop for yourself based on what the chickens like to eat in leftovers!

Selecting a Breed

There is much discussion within the poultry community as to which breed is better, with each one having its own set of attributes. There is no real right answer to the question of which breed of chicken is the best, but there are certain aspects of each breed that make them more or less desirable for the backyard flock. This is determined by what are you looking to get out of your birds and how much experience you have at raising them.

Before ever getting your first bird, there are many questions that you must ask yourself. Do I want eggs or meat, or maybe both? Are my birds going to be free range or are they going to be kept within a coop? What about interaction with the kids or grandkids? Do I have to worry about weather extremes? How much space can I devote to raising chickens and how much time do I have? The list of questions can be endless, but there always seems to be a breed that will fit your needs.

Let's take a look at some of the more popular breeds that are kept in backyard flocks.

There is a wide variety of poultry combs.

RHODE ISLAND RED

This is the quintessential breed that almost every new chicken owner seeks to have because that is what they remember Mom and Dad or Grandma and Grandpa having when they were young. The Rhode Island Red is a large breed of chicken with the hens averaging six and a half pounds and the roosters averaging around eight and a half pounds. They are great layers of large to extra large brown eggs. They are a great dual-purpose bird, meaning they can be used for both eggs and meat. The hens on occasion will go broody and so will naturally expand or replenish the flock. They can be free-range or be adaptable to life in a coop. They are a great all-around breed. There is, however, a "but" . . .

Rhode Island Red roosters can be quite intimidating, and, depending on your level of knowledge about caring for a backyard flock, can become quite controlling of you and your family. Many stories have been told of how Grandma's old red rooster used to chase the kids around the farm and try to strike at any chance. This is one breed that you have to remain the top of the pecking order with or you will have problems and your chicken-raising experience will quickly become a bad one. I cannot remember a time when Rhode Island Reds were not part of my flock and

Rhode Island Reds are great dual-purpose birds. They can be used for eggs and meat.

at no time was a rooster ever allowed to dominate me. As long as you stay on top of them, this is a great breed to have.

NEW HAMPSHIRE RED

If you are new to chicken raising but you want a red chicken like Grandma and Grandpa used to have, I suggest starting out with New Hampshire Reds. This breed was developed strictly from the Rhode Island Red. It was carefully bred for temperament as well as all the same attributes as the Rhode Island Red with the great egg laying ability and nice meat production. Body weights are about equal, with the roosters being slightly smaller in stature. The New Hampshire breed has not been as overbred as the Rhode Island and, therefore, tends to be a bit more broody and can often supply you with a good number of chicks with which to propagate your flock. They are much easier to handle and are not nearly as stubborn to learning as the Rhode Island Red. The New Hampshire Red is a great breed of bird for those with little or no experience, and you will be rewarded with ease of care, great egg laying and the occasional fryer, if you so choose.

BARRED PLYMOUTH ROCK

This breed should not be confused with the Dominique breed from which it originated. The Barred Plymouth Rock—or Barred Rock, as it is commonly called—is another large breed of brown egg laying variety. This breed is somewhat larger than the Rhode Island or New Hampshire Reds with the roosters typically weighing in around nine and a half pounds and the hens around seven and a half pounds. They are best known for their striking black and white barring in their feathers. Even though they are a large bird, their temperament seems to be quite docile. The hens tend to be a bit more broody than the Reds and they make great mothers. The Barred Rock can be used either as a free-range bird or kept in a coop providing that it is large enough that the birds are not crowded due to their size. They are a great dual-purpose bird. The Barred Rock is a good choice of breed if you are going to have children around the chickens, as long as it is established early on who is the boss and who is a chicken. They are a relatively fast growing bird and will generally start laying the first eggs around the eighteenth to twentieth week.

Barred Plymouth Rocks are fairly docile birds and well-known for their distinguishing black and white barred feathers.

Delawares have a mild temperament and are generally great around children.

DELAWARE

This is by far my favorite of all the heavy breed chickens. The Delaware was the original dual-purpose bird when the commercial meat and egg revolution took off in the 1950s and '60s before they switched to genetic mutations of Leghorn and Cornish breeds. Delawares are a large bird with the males averaging eight and a half pounds and the females around six and a half pounds. They have a mild temperament and are generally great around children. The hens lay a large to extra large brown egg and the males make great table fare. The hens tend to mature quickly and will start laying around eighteen weeks of age. Depending on the breeding, the hens could very well go broody and raise their own young, and both sexes make doting parents.

Good quality Delaware chicks may be difficult to find but are well worth the search. I have found with my own Delawares that they are very nurturing of other flock members and I use them exclusively as buddies for injured or disabled flock mates. I have first hand witnessed one of my young Delaware cockerels take care of a disabled Ameraucana rooster who was progressively going blind. The Delaware would nudge the Ameraucana toward food and water, show it where to eat and drink, and then stay with it for hours to protect it from the other members of the flock.

The Delaware rooster is slow to anger and tends not to want to fight with other roosters within a flock, but if needed it will fiercely protect his girls from predators. They are quite romantic and a good rooster may find himself in love with the farmer's wife, as has been the case here on the farm with a few generations of boys. The hens tend to be a little more standoffish than the roosters. If you are looking for a great starter bird with all the right qualities, you couldn't go wrong with a Delaware.

These four breeds would have to be my top picks for a backyard flock of brown egg laying, nice meat producing birds. But what if you want something a little different? My top choice would have to be the Ameraucana. There is great confusion in the term Ameraucana and many newcomers to the chicken world find themselves on the wrong end of the breeding stick when it comes to this breed.

AMERAUCANA

There are three closely related breeds: the Araucana, the Ameraucana and the Easter Egger. To make things as simple as possible, the Araucana is a rump-less chicken that lays a blue egg. These birds also have distinctive tufts of feathers that protrude at the ears and a pea comb. The Araucana was cross-bred to create the Ameraucana. The Ameraucana has a tail, ear muffs, a beard, a pea comb, and lays a blue egg. Some versions of what is called an Ameraucana will lay a green egg. The Easter Egger is an Ameraucana that has been cross-bred, lays various color eggs, may or may not have ear muffs and a beard, and could

The Ameraucana has a tail, ear muffs, a beard, a pea comb, and lays a blue egg.

have a single comb or a pea comb. The Easter Egger is basically just a mutt chicken with a fancy name and pretty eggs.

The Ameraucana as we know it is a great bird that lays a medium to large egg. True Ameraucanas tend to be on the smaller side, with the roosters averaging around six and a half pounds and the hens around five and a half pounds. I have always said that Ameraucanas are the Siamese cat of the chicken world. They project the attitude that, "I will come near you but I will only let you touch me if I want to be touched." I have a couple of them that love to be held once you get ahold of them, and I have others that will come within inches of your hands but run away if you try to touch them. They are a great novelty bird, but at the same time can provide you with plenty of eggs and the occasional meal.

LEGHORN

If you have only ever eaten store-bought eggs and have a real aversion to eating anything other than a white egg, then your breed choices are very limited. The best white egg-laying breed would be the Leghorn (pronounced Leg-urn). The Leghorn is a profuse layer of white eggs, but there are many drawbacks to the breed, especially for first-time chicken folks. They are on the smaller side, with the roosters going about six pounds and the hens around four and a half pounds. Leghorns are a flighty breed and tend to be quite noisy; therefore, they are not well-suited for living close to neighbors. The hens will rarely set, so it is not unusual to have to buy new chicks each time you want to expand your flock or incubate the eggs yourself. If you are looking strictly for an egg-laying chicken, the Leghorn may be just what you are looking for. Be forewarned that they are not generally a sociable chicken nor are they particularly good around children, but we must give credit where credit is due by saying they are probably the best egg-laying breed out there.

Did you know?

The world record for the most eggs laid was set in 1979 by a White Leghorn who laid 371 eggs in 364 days.

SEX-LINKS

In the quest to find the right chicken for your needs, you will undoubtedly come across the term "sex-link." A sex-link chicken is a hybrid created by crossing two different breeds of chickens to create a chick that is gender sex-able at hatch. There are a wide variety of sex-link breeds, with the two most common being the Black Sex-Link and the Red Sex-Link. Both are equally good at laying large to extra large brown eggs. Because they are sex-able at hatch, this makes a great choice if you only want or can only have female chickens where you live. They have a nice body size and usually an even temperament. The one drawback to sex-link chickens is that they are single generation birds, meaning that you cannot breed a sex-link to a sex-link and get a sex-link chick. Any resulting offspring will revert back to the parent stock, which is to say if you were to breed a Black Sex-Link to a Black Sex-Link, the resulting chicks would look more Barred Rock than anything.

Sex-links tend to start laying earlier, usually around the seventeen to eighteen week age range, and lay very well for about the first year and a half, then taper off fairly rapidly. In contrast, a standard breed will lay well for the first two to three years, then taper off more slowly. It is a great breed to have if you have children. The roosters make fine roasters and can be purchased as chicks at a real cut rate from the hatcheries.

CORNISH CROSS

But what if you are only looking for a chicken that will fill your freezer with meat each year? There are a few different varieties out there and the hatcheries seem to come up with a new hybrid just about every year that is supposed to be the latest and greatest thing since KFC. It has been for many years now that the Cornish Cross was the go-to bird for meat production. In a backyard setting, they are usually ready to butcher in fourteen to eighteen weeks at about the five-pound size. These birds eat massive amounts of feed, grow fast and have many health problems due to their fast rate of growth. These birds are kept in smaller pens to limit their movement and are just allowed to grow.

Newer on the market is what is termed as Freedom Rangers, which are a fast-growing meat-type bird, but they have the ability to free range while still putting on body weight at a more rapid pace. They will generally be ready for slaughter around twenty-four to twenty-eight weeks of age. They don't have nearly as many health issues as the Cornish Cross. The one drawback to the Freedom Rangers is the look of the carcass at butcher. The Cornish Cross is a white-feathered bird with light-colored skin, and so a dressed bird looks like you would get from the store. The Freedom Rangers are generally a red-feathered bird, and so the carcass will show some feather follicle discoloration at butcher. This has no reflection on taste, just on the presentation of the bird on the table.

If you should decide that you would like to try your hand at raising your own meat birds, keep in mind the amount of feed that they will consume. With today's feed prices, it would not be unreasonable to have to invest around $18 per bird in feed to get them to market weight. This cost could be reduced if you have access to a bulk feed mill or if you choose to go with one of the newer hybrids that can be put out on pasture and grow at a slower rate.

There is a myriad of breeds, hybrids and color variations out there and each one has its pros and cons. A good starting point when choosing a breed would be to take a look at Henderson's Handy Dandy Chicken Chart, available online at www.sage-henfarmlodi.com/chooks/chooks.html. This gives a comparative listing of about sixty common breeds as size, hardiness, egg production and more. Also talk with other chicken enthusiasts in your area to see what breeds they have had the best luck with, and if you have the opportunity, try to visit a few backyard farms in your area to get a sense of what different breeds are like in their own setting. There will always be good and bad with each breed, some more than others. Don't be afraid to try various breeds until you find the one that suits you right. We all have our favorites and those listed above just happen to be mine, as well as the ones that I have had the best luck with. Over the years I have tried many different breeds and at one time kept about twenty-eight different breeds, but I found myself leaning more and more to just a handful of breeds that suited my situation the best. Each person's abilities and style of raising is different, but there is a breed out there to suit each one.

Did you know?

The first American poultry show was held on November 15–16, 1849, in Boston. There were 1,423 birds shown by 219 exhibitors. At that time there was no standard established for each breed. The American Poultry Association was formed in 1873, and the next year published the first American Standard of Excellence.

Building a Coop

I do not want to spend too much time on building a chicken coop simply because that is a book in itself and there are lots of books out there on this very subject, but I do want to touch on a few points that come to mind.

A coop is limited only by your ability, your imagination and your wallet. But then there is also the reality of it being a chicken coop. Yes, it is nice to look through different farm magazines and see these extravagant coops that people have built for their chickens and dream that some day your flock will have a better and bigger one. But why? A chicken coop needs to serve the purpose of shelter, a roosting place and a place for the hens to lay their eggs. I know it would be cute to have a little gingerbread house with a little white picket fence around it and a nice little play yard for the birds, but those magazines never show you the after pictures from two weeks later when the chickens have eaten all the landscape plants, kicked up dirt on all the nice paint from scratching around, and they created all that dust that the spiders just seem to love.

A coop is utilitarian. Yes, you want to have the nicest one that you can afford to build, but stay realistic. A coop needs to provide shelter from both the weather and predators. It should be raised off the ground so to avoid ground-dwelling animals. It needs to have a good roof to protect the birds from rain, snow and sun. It needs to have secure walls with ventilation to provide good airflow, yet not be drafty. And it needs to have a secure door so that the birds can be locked in and the predators locked out. It needs to be of sufficient size for the number of birds that you are planning to keep. As a general rule, you want to figure a minimum of two square feet per bird if you are going to have an attached outdoor run. You will need to figure a minimum of five square feet per bird if they are going to be housed full time in the coop. The more space that you can provide for your birds, the happier they will be and also you will be. Cramped birds cause problems.

Your coop will also need to provide space for the chickens to roost at night. Your roosts should be no higher than thirty-six inches off the ground, preferably about twenty-four inches, and provide about one foot of roosting space per bird. The roost itself can be made out of just about anything that is about one inch in diameter. Wood is preferable, as it is easy for the birds to grip with their feet.

Your coop also needs to have nesting boxes for the hens. You should provide one box for every five to six hens. The boxes can be mounted to the outside of the coop to conserve space, and have access from the inside of the coop. The nesting boxes should be about fourteen inches square and about twelve inches high. Many people use milk crates for this purpose if there is room inside the coop for them. Nest boxes can be at floor level or elevated on the wall. If you are going to elevate them, do not place them more than about twenty-four inches off the floor so that the hens can easily access them.

Your coop needs to be equipped with a door that securely latches. Many a chicken has been lost to a raccoon that has opened the door. The door needs to be large enough to allow you access for cleaning purposes. Many people will also build a pop door that is just large enough for the chickens to come and go freely. This is especially helpful in the northern climates where you will need to keep as much heat inside as possible and as many drafts as possible outside.

The roof of your coop should be suited for your climate. If you have to deal with snow, then your roof should be solid built to handle the weight of your typical snowfall. There is nothing worse than waking up on a cold winter's day to find that the snow that fell overnight has collapsed the roof down on your birds. If you live in a rainy area, the roof needs to overhang enough to shed the water away from the coop. If you have to deal with the beating sun during the summer months, then you will want a roof that is reflective so that it does not gather too much heat and make the inside of the coop any hotter than it needs to be.

That brings us to ventilation. Your coop has to be ventilated. It needs to provide plenty of airflow without being drafty. Ammonia will build up inside your coop from the droppings, and if left, can cause serious respiratory issues. The ventilation should be covered with strong wire to keep predators out as well as louvered grates to keep the weather out. I don't know how many times I have seen coops filled with snow and water that has come in through the ventilation openings.

You can also insulate your coop. This is helpful for the folks that live in the northern states. Winters can be brutal on you and your birds. Insulated floor, walls and ceilings help protect your birds from the cold. Any type of insulation can be used, but it will have to be covered or the chickens will peck and scratch it apart. They especially enjoy Styrofoam and make a huge mess in a short time, let alone what they may eat of it.

If you are going to provide your birds with an outdoor run, it also must be very secure from predators. Provide about ten square feet of space per bird for their outdoor run. Part or all of it should be covered so as to provide additional protection from the elements. The perimeter of the run should have some type of protection against digging animals. Cover the exposed areas of the top with netting to keep your birds in and flying predators out. You should also have a door on your run so that you might access it if need be.

I know, you are reading through this and thinking, "Wow, this is going to cost a fortune to build a coop that has all these features." Not really. With a bit of patience and ingenuity along with some scrounging, you can probably build most of it for free. It might not be the fancy little gingerbread cottage you saw in the magazine, but it can come pretty darn close.

Start by asking your friends and neighbors for any unwanted building materials. If there is any construction going on around you, visit the site supervisor to see about getting discarded materials. A great place to get dimensional lumber and plywood is a diesel repair shop, the places that work on the big tractor-trailer rigs. Most of their parts come in on skids or in crates, which they generally break down and throw away. If you ask the shop manager, they will usually be happy for you to take them. Another good source of lumber is used pallets. These can be pried apart and used in a variety of different ways. When pulling apart skids or pallets, try to save the nails as you can certainly reuse them. Look around your area for people tearing down old sheds and outbuildings. There is usually enough good lumber left in them to justify offering to tear it down for them if you can have the materials.

Roofing is usually a big expense, but it does not have to be. Roofing can be obtained in the same way as other materials. Visit home construction sites for leftover roofing. Visit your local hardware stores for broken bundles or damaged roofing. Check with the local billboard companies to see if you can get discarded old billboard tarps. Check with the neighbor about his old metal shed out back. Yes, the roof might leak, but the walls are usually good enough to use as nice metal roofing.

Your roosts can be a 2 × 4 that you split down the middle and round off to make it like a dowel. Old shovel, rake or broom handles also do the job well. A sturdy tree branch can also be used.

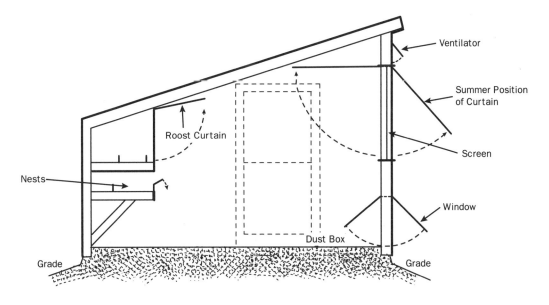

Diagram of a chicken coop

For your nest boxes, use either milk crates or produce crates if you don't have the materials to build nest boxes. Bedding for the nest boxes can usually be had for free if you visit a local hay supplier and pick up the loose hay that has come off the bails.

Screen and wire are going to be your hardest materials to come by, but again, check with neighbors and friends. Maybe someone has an old coop or pen that they want taken down, or check with local farmers to see if they have wire mesh fencing that they want removed. It is out there, but you have to look.

You can build a nice coop and run out of scavenged materials if you are not in a huge hurry. If you make it a priority and also a game to find suitable materials, you will be surprised at what you will be able to find. You might have to buy a bit of hardware, such as nails and hinges, but the rest of the materials you should be able to get for free or at little cost. And a coop does not have to be built out of lumber. It can be built from stacked hay bails, an old travel trailer or whatever your imagination leads you to use. As long as it is suitable for your climate and provides your birds with protection, then it will work. Have fun with this project.

Nest Boxes

A good nest box should be your hen's best friend. This is where she will want to return to nearly every day to lay her egg. But what if she doesn't want to lay in the box that you have provided for her? A hen has to feel comfortable with her surroundings in order for her to lay an egg, and this is very evident in the place that she chooses to lay. You can build the fanciest, nicest nest boxes around, but if the hen does not feel safe and secure in them, then she would rather lay her egg on the floor than inside the box.

Just about anything can be used for a nest box, whether it is a plastic milk crate or a fancy commercial nesting system. But what matters most is that the hen feels comfortable. For years, we used plastic produce crates with the fronts cut out of them, stacked and zip-tied together. These worked fine, but we found that the girls wanted to roost on them and therefore would poop in the nests. We went to using plastic totes with a hole cut in the front and found our girls really like them. They provide a secure area for the girls to lay and two hens can easily crowd inside one box. As a general rule, you should provide one nest box for every five to six hens. But there is a paradox here, that even if you only have a few hens, you should still provide at least two or three boxes so that they can choose which one they like best. And just because they like one this week doesn't mean that they will like it next week.

The position of your box is key. Whatever your nest box situation is, if the girls do not like to use them, then the answer may be as simple as changing the position of the box. If left to her own devices, a hen will find the most out-of-the-way, secluded place in which to build a nest to lay her eggs. This should also hold true for their nest boxes. A darkened corner of the coop or a separate nesting area seems to work well.

Nesting materials can also vary greatly. Most people will use straw or hay in the nest boxes. Some will use pine shavings and others will use shelf mat liner. Whatever material you choose to use should be dry, clean and easy to clean or replace.

If you have young pullets that are just getting to laying age, it is advised to place golf balls or other suitable fake eggs in a nest or two to show the girls that that is where they are to lay. Don't be surprised if you have ten nesting boxes for twenty hens and they all want to use the same box. This is very common. In our breeding

Almost anything can become a nest box—plastic milk crates to commercial nesting systems. Just make sure the hen feels comfortable.

If you are going to have only one level of nest boxes, place the bottom of the box approximately sixteen inches off the ground. This height discourages ground predators from getting into the nests, but don't position them so high that it makes it difficult for the hens to get into.

A nest box for a standard hen should measure a minimum of 12"x12"x12". Keep in mind that, on many occasions, more than one hen may be using the box at a time. It is always better to go bigger than to go smaller.

pens, we supply one nest box for every six to eight hens. If the nest box is occupied, the other hens will line up and wait, or if they can't wait any longer they will simply double up.

A nest box for a standard hen should measure a minimum of 12" × 12" × 12". A large hen would require more space and likewise a smaller hen would require less space, but keep in mind that, on many occasions, more than one hen may be using the box at a time. It is always better to go bigger than to go smaller.

If you are planning on letting any broody hens hatch out their own eggs, then you will want to plan for this when establishing the nesting area. In our main hen house, the nesting boxes are of the produce crate variety. The bottom boxes are turned upside down. This gives a floor level nest that the broody hens like to use so that the chicks can easily get in and out of the nest. This also allows us to place a bottomless cage around the nest to help protect the new chicks from getting trampled in the first few days until they can get up and travel with the mama hen easily.

Your top boxes should have a sloped top on them to help discourage roosting and pooping on top of the boxes. A 30-degree angle or better seems to work well as it is too steep for them to stand on.

If you are going to have only one level of nest boxes, place the bottom of the box approximately sixteen inches off the ground. Granted, you can place them as low as you want, but this height helps discourage ground predators from getting into the nests. You don't want to place your boxes too high either as it makes it more difficult for your hens to get into. Your top box should be placed no higher than three feet off the ground. Some folks will build their nest boxes with a roost in front of the box opening, thinking that this will help the hens when they are trying to enter the box. All this really does is promote roosting. If the box opening is sufficiently large enough for the girls to get into, then they do not need an added roost.

It doesn't matter what you use for a nesting box, what matters most is location, location, location. If you find that your girls are wanting to lay most of their eggs on the ground outside of the nests, then they are telling you that they do not like the nest position or the size of the nest. Also, be sure to clean out the nest boxes regularly and spray them with a good pesticide before placing new nesting materials back in, as nest boxes are a favorite hiding spot for parasites. Keep your girls happy with clean, dark, secure nest boxes and they will return the favor with their eggs.

Artificial Lighting

There is some controversy over the use of artificial lighting to keep hens laying through the colder, darker months of winter. Some say that this is nature's time for the hens to rest and get rejuvenated for the spring ahead. Others say that the hens need to keep laying so that they are not just freeloading through the winter. I feel that both sides of the argument are correct but I tend to lean more to the side of keeping them laying.

The use of artificial lighting through the winter months will help the girls keep laying, and if done correctly, can help keep them healthier as well. The drawback to using artificial lighting would be if you have roosters, the neighbors might not appreciate them crowing at 4 A.M., but then if your roosters are like anybody else's, they crow all night anyway.

A lighting program can be as simple or as complex as you wish to make it. I personally like the simple approach. A laying hen requires a minimum of fourteen hours of light per day to remain on a normal laying schedule. During the winter months, daylight hours may drop to as low as seven or eight, or if you live in the extreme parts of the world, it could be even less than that. With the use of lights and a timer you can create sufficient light hours to trick your girls into laying.

Fall and winter are tricky months for chickens. Starting into fall, your birds will be going into molt, loosing their feathers and egg production will slow or stop altogether. This is a time of rejuvenation for the birds, more so for the hens than the roosters. In the large commercial egg producing plants, they can force molt the hens at any time during the year by the regulation of their lighting and feed schedule. Because the large poultry houses are closed off to the outside world and the hens have no idea if it is day or night, hot or cold, this season or that, their natural cycles can be manipulated in many different ways. For the backyard flock raisers, this is not as easily accomplished.

Molting is a necessary function of nature that allows the chickens to put on a new, heavier coat of feathers to help protect them through the cold winter months and prepares the hens for the upcoming spring and chick hatching. The theory of molting is covered in another section. Once the molt is complete, your daylight hours have dropped considerably and the result is that the hens will generally not start laying again until spring. This does not have to be the case. Through artificial lighting, you can get the girls back up to laying right through the winter.

Let's digress a bit here. Let's say that you get your chicks in the early spring. You raise them up all spring and summer. Around August they start to lay. The fall equinox is around September 22, and so the days begin to get shorter. Your birds will go through a molt and then be plunged into the darkness of winter. After the winter solstice (December 21), the days begin to get longer. By early spring there is enough light to coax the girls out of slumber and they slowly start to lay again. If you have hens that like to go broody, then they will want to set on eggs for the spring hatch. Three weeks of setting followed by six weeks of chick rearing and then the girls are finally back to wanting to lay eggs. That is a really long time to be feeding all those birds with very little in return.

Now let's look at it from the other perspective. You get your chicks in the spring, raise them through the spring and summer, and they start to lay around August. Then they go into their molt. You feed them higher protein foods to help quicken the pace of the molt. At this time you start with an artificial lighting program to slowly get the light hours back up to fourteen hours or greater. The girls' natural instincts tell them that spring is coming and so their bodies prepare for egg laying. As the molt ends, your lighting schedule is close to that magical fourteen-hour limit. Their minds are telling them that it is cold and dreary outside but their bodies are telling them that spring is here and so it is time to lay. They begin laying, and by not allowing them to go broody, you can have eggs right through winter. As spring approaches, you lessen the amount of time the artificial lighting is used until you get back to natural lighting and the hens keep right on laying. It is a lot easier on you to see some return on your feed investment and the girls are still just as content as ever.

As stated in the beginning, your lighting program can be as easy or as complex as you wish to make it. I personally like the easy approach. We have a large flock of chickens by backyard standards. We have separate pens that are all conjoined that house all our breeder birds. We also have a large main pen area that houses all of our younger birds and also our production birds. It is roughly split 50/50 with the numbers of birds in the breeding pens versus the number of birds in the production pen. During the darker months of fall and winter, we use our breeder birds for egg production, and let our production birds rest through the winter. Yes, we will continue to get a limited number of eggs from our production birds, but not nearly the numbers we would normally get.

We run a string of small Christmas lights down through all the breeder pens; the lights are attached to a simple timer. As the girls are just starting to come out of molt, we time the lights to come on earlier and earlier each week. At this time of year, we are normally at about twelve hours of natural light. Dawn starts around 7 A.M. The first week we will time the lights to come on around 6:30 A.M., and go off at 7:30 A.M. Each week we turn the lights on a half hour earlier until we get to where they turn on at 4 A.M., and by then they are turning off around 8:00 A.M. Combined with the natural light, this gets the girls on a fourteen-hour light schedule.

You may think, why not increase the lighting in the evening hours so that way the roosters are not up crowing in the wee hours of the morning? Well, chickens do not see well in the dark. If you were to run your artificial lighting to increase the length of the evening hours, then when the timer clicks off the chickens are plunged into darkness and have problems finding their roosts. This can lead to unnecessary injury and trauma to the birds. It is always best to allow the birds to go to roost under natural conditions.

But then you ask, why do we not run artificial lighting for our production birds? After all, that is what they are for. Our production birds are in a large roosting area at night. They have a separate hen house for laying and they have a large common yard for daytime activity. It would take a lot of lighting to be able to light up all these areas sufficiently to get the girls to continue to lay. It also opens up the possibility of predation by owls and other nocturnal animals. Our breeder pens are fully enclosed with wire, whereas our production pen is fully open, except for the roosting house. If we were to run lights for our production birds, that would mean that I would have to be out every morning at 4 A.M. to let them out, whereas the breeder birds can just go about their day until I go out at first natural light. This is what works very well for us.

If you have a smaller scale coop with an attached wire run, then you can certainly light up the whole area. Place a small light in the coop and a small light in the run and you would have sufficient lighting to coax the girls into continuing to lay.

The lighting itself does not have to be anything significant. You are not trying to give the birds a suntan. The way to figure if they have sufficient light is, after your eyes adjust to the darkness, if you can see the level of water in one of the waterers made of opaque white plastic from eight feet away, then you have enough light. To me that is a bit too dark. Our breeder pens are approximately five feet wide and sixteen feet long. One 7-watt nightlight bulb in each pen at ceiling height provides enough light to get the birds up and moving.

We prefer to use colored lights in our pens. Research has shown that red light is good and blue light is better for production. With our string of Christmas lights, we have a combination of colors. This also makes it not quite as bright as clear bulbs would be for the neighbors' sake. Plus it looks more festive. Our whole lighting program cost us about $12 to set up. It takes minimal power to operate it through the winter and we gain significant egg production to help offset the cost of feed. Our birds are happy and healthy and our neighbors don't mind because they get supplied with fresh eggs through the winter.

The other aspect of running artificial lighting that I have touched on is the health aspect. We live in the South, so our winters do not get bitterly cold like in the North, but we still see weeks with temperatures from 10° to 20°F (-12° to -6°C). Chickens produce heat from feed. In the colder northern regions the risk of exposure to the elements for longer periods of time can cause severe problems with birds. If your chickens are roosting for twelve to fourteen hours every night, that is time that they can be overexposed to the severity of the weather. A chicken will eat large amounts of feed just prior to going to roost in the evening. This is to allow their bodies to continue to produce heat through the night until they rise again in the morning to eat again. By using artificial lighting, this allows the birds to rise earlier and replenish their feed to better regulate their body temperature.

The coldest part of the night is just before dawn. If you can have your birds up and eating and moving prior to this time of day, they will be better prepared to handle the extremes. Because of this, they do not get as easily chilled and, therefore, can better fight off ailments.

So all in all, it is beneficial to you and your birds to put them on an artificial lighting program. You will have more eggs, better use of feed and happier, healthier birds because of it.

You know you're addicted to chickens when . . . you spend more time cleaning the girls' coop and run than you do cleaning the house.

Weather Extremes

Nearly every nation in the world has chickens in one form or another. From the blistering sands of Egypt to the frozen tundra of Siberia, chickens are thriving in extreme heat and cold, but each year we get numerous e-mails wanting to know why someone's chickens froze to death or died from heat exposure and what can be done to prevent it.

Chickens are adaptable animals to some extent, but it has a lot to do with the breed that you choose and how they are kept. Yes, it is possible to keep heavy bodied, heavily feathered Cochins in the heat and humidity of south Florida. And yes, it is possible to keep light bodied, large comb and wattle Leghorns in Alaska, but neither is recommended. Certain breeds thrive better in certain conditions, depending on their body structure, feather coverage and comb and wattle size. The Orloff in Russia and Siberia, the Malay in Asia, or the Turken in Hungary, each breed was originally bred for the surrounding area, but breeds have been moved around the world and have become adaptable to some degree. If you want to keep Leghorns in Alaska or Cochins in Florida, then be prepared for a bit of extra work.

The biggest problem that we see is birds that freeze to death, but we also see where many a live bird has been buried just because someone thought it was dead. A chicken's body temperature can vary widely. Normal body temperature runs between 105ºF (41ºC) and 107ºF (42ºC). The low lethal limit body temperature is around 73ºF (23ºC) and could run lower in certain circumstances. The upper lethal limit body temperature is between 113ºF (45ºC) and 117ºF (47ºC). That is a really wide temperature variation, so don't ever consider your ailing chicken out of the picture until you make sure.

People feel that if they are cold, then their birds must also be cold. That is not always the case. Birds, if left to their own devices, will grow in a nice layer of downy feathers prior to winter. This down is covered by a thick, tight layer of main feathers. But where the problem arises is when the temperature drops to about 30ºF (-1ºC), and people put their birds under artificial heat.

Let's say you live in Florida. During the summer you can have extended periods of temperatures in the 90ºF (32ºC) range. All of a sudden a cold front comes through and the temperature drops to 70ºF (21ºC). You would be wearing a coat to stay warm. But say you live in Maine instead where your temperatures have been in the 50s, and you take a trip to Florida during that 70ºF (21ºC) "cold" snap, and you would be putting on shorts and a T-shirt to try to stay cool. It is all because you have become acclimated to your area and its temperature.

The same holds true with chickens. If they are of a breed that is bred for the conditions and are properly acclimated, then you should be faced with much fewer problems. A fully feathered chick at six weeks of age can handle temperatures in the 20ºF (-7ºC) range with dips below zero, but what happens many times is that people who artificially brood their chicks will keep them under heat for six weeks and then put them outside in the cold and expect them to live. The chicks have not been acclimated to the outside temperatures and, therefore, have not developed enough down to keep them warm. This holds true with older birds as well. If your fall and winter temperatures are getting down into the 30s and you start running heat in the coop for them, they will not develop the feathering they need to keep them warm. They will only develop feathering to keep them warm at 30°F (-1°C).

I have read many stories from people who live in extreme cold conditions such as Alaska, Canada and Northern Europe, who have healthy, happy chickens in sub-arctic temperatures of -20°F (-29°C) and -30°F (-34°C) and no artificial heat source, but they have properly prepared their chickens by letting them acclimate themselves to the temperatures and also by providing the birds with proper shelter. You are not going to be able to take a Leghorn, put it in a wire cage, drape a tarp over the cage and expect the bird to survive in -30°F (-34°C) temperatures.

If you are fortunate enough to have a broody hen that hatches out a clutch of chicks naturally, take some time to observe their actions. It doesn't matter what the outside temperature is, the hen will have the chicks out of the nest at about three days old, teaching them to peck and scratch. When they get cold, they will run under the hen to warm up and then right back out. The older they get, the longer they can stay out from under the hen because their feathering is coming in the right way and they are getting acclimated to the outside air.

The standard for artificially brooded chicks is that you start off at 95°F (35°C) temperature for the first week, and then drop the temperature by five degrees in each following week. Chicks are normally fully feathered by six weeks of age. That would mean that if you kept them under heat for the six weeks, by the time they are feathered and to be transferred to a grow-out pen, the temperature would be at 70°F (21°C). Well, what if it is only 40°F (4°C) outside? The chicks will not be able to readily handle the 30-degree temperature difference and will be chilled, causing all sorts of health issues and possibly even death.

Granted, most chicks in the U.S. are hatched in the spring as temperatures start to rise. Here in the South, we start hatching our chicks in early October so that our pullets are ready to lay by spring when the rest of the folks are just starting out with chicks. We hatch non-stop from October until June and our chicks hit the grow-out pens at three weeks of age or younger. We do this by accelerating the temperature drop during brooding. Instead of dropping the temperature 5 degrees each week, we drop ours to whatever the chicks can comfortably handle. For the first two to three weeks, they are brooded in our office under an 85-watt red floodlight. We watch the chicks and raise the light as much as possible to just where they will not huddle together. By three weeks of age, the temperature is at about 70°F (21°C). The chicks are then moved to a grow-out pen that is weather protected, and a 125-watt brooder lamp or an 85-watt floodlight, depending on weather conditions, is placed over them. Again, we raise the lamp to just where they are comfortable. By week four, we have switched back to the 85-watt floodlight. By week five, they have light only at night, and by week six, they are fully acclimated to the outside temperature. Yes, this is the deep South, but it is not unusual for us to have temperatures into the teens.

Your biggest concern should not be temperature as much as drafts. Chickens can handle much colder temperatures if there are no drafts. Wind chills affect poultry just as they affect us. Be in a room that is 30°F (-1°C) and it is cold but tolerable, if you are dressed for it. Now step outside where it is 30°F (-1°C) but there is a 10-mile-per-hour wind blowing. It now feels like it is 16°F (-9°C). Therefore, it is very important for your chicks or chickens to be kept in a draft-free area. Chickens produce heat and if you have a coop that is well protected from drafts, even though it is 30°F (-1°C) outside and blowing 30 miles per hour, just from the chickens' own body heat, it may be 40°F (4°C) inside the coop.

This also brings up another point. If you are going to keep just a few birds and you live in a cold weather area, they do not need a roosting area or coop that is the size of your house. The smaller the area, the easier it will be to keep warm even with just their own body heat.

But when do you need to add artificial heat to your coop? In all actuality, you really should never have to heat your coop if it is properly constructed and your birds are of a breed suited for the cold temperatures. About the only time that this would differ is if you have young birds, birds with health issues or birds that have not yet completed their molt before the cold weather sets in.

I have read many stories of chicken coops and barns burning to the ground because of artificial heat being used to keep the birds warm. Chickens only need to be made comfortable, not placed in a sauna. Naturally, if you place a 250-watt heat lamp bulb a foot off the floor of your coop in the middle of winter, where do you think your birds are going to hang out? Right around that lamp. Chickens are clumsy and also inquisitive. They will bump into a low-hanging lamp or try to roost on it or the cord. The lamp can easily be knocked to the floor. Or, if you have litter on your coop floor such as pine shavings or straw, the chickens are going to naturally scratch around in it and can kick it up onto the lamp. Both of these could cause a fire.

Think of it this way, if you have a well that you get your water from, during the winter most people will put a small light inside the well house to keep it just warm enough for the water not to freeze. You wouldn't put a space heater in the well house; otherwise you would have hot water coming through your cold water lines. The same is true for your chicken coop. Use only enough artificial heat to keep them from freezing, not cook them.

Another good thing to remember with the onset of winter is that protein equals heat. The higher the level of protein in your birds' diet, the more body heat they are able to produce. If your birds have been out free ranging all summer, then their bodies will be leaner from the added exercise. As the weather starts to cool, keep your birds cooped in a pen longer and feed them higher protein foods to help fatten them up. Added corn and wheat products help fatten them. If you can't find high-protein chicken feed in your area, get a bag of kitten kibble and mix it in with their feed. By springtime you will have a bunch of fat, sassy hens that stayed healthy through the winter because of the added protein in their diets.

OK, so let's say one day the birds are all out free ranging when a great nor'easter blows in. Temperatures drop to near zero, white out conditions, snow is piling up fast. The chickens all run for the safety of the coop. By the time you are able to get bundled up and out there to shut them in, there is a foot of snow on the ground. You race out to the coop and along the way you find Miss Cluckers buried in the snow. She is lying there lifeless and you think only the worst. She couldn't make it back to the coop in time and got lost in the snowstorm and has frozen to death. The tears freeze on your cheeks. Take a deep breath. Pick up Miss Cluckers, straighten out her feathers, get her into a normal sitting position and stick her inside your jacket. Close up your coop to keep the others safe, and take Miss Cluckers into the house. Wrap her in a warm towel or blanket and place her in a warm area. Massage her body a bit and watch for any breathing or eye movement. Usually within a few minutes, signs of life will reappear, and soon Miss Cluckers will be up and running around like nothing ever happened. Yes, this is very dependent on length of exposure, but never count a chicken out until you make sure.

There are a few preventative measures that can be taken if you will be experiencing a very cold night. Besides providing them with adequate shelter, make sure the birds have plenty of good feed and water prior to nightfall, and when the birds go to roost, take Vaseline petroleum jelly or Vicks VapoRub and smear a light coating on their combs, wattles and feet. This will help retain some body heat and help prevent frostbite on those extremities.

The other side of the spectrum is heat. Heat is much harder to deal with than the cold. Just as as you or me, if it is cold we can put on extra layers to stay warm. But if it is hot, you just can't take off enough layers. Now think of wearing a down coat on the beaches of the Bahamas.

Birds do not sweat. They cool themselves by raising their wings to allow more airflow next to the body and by panting to cool their core temperature. Chickens have many air sacs within their bodies. They breathe in cool air and breathe out hot air. This helps them to regulate their body temperature. When the weather gets extremely hot, their cooling systems cannot keep up with heat exchange.

It is very important to provide your birds with suitable shelter against the heat just as it is against the cold. If you live in an area such as the desert southwest or a point even farther south, you will want to take the opposite approach to those who live in the Great White North. Your coop will need to provide shade as well as great ventilation. You would not want to build your coop entirely out of metal because it would become an oven in the heat. In the North, you could get away with having a smaller coop because the birds will be able to provide enough body heat to keep it warm. In the South, use a bit larger coop so the birds can spread out and dissipate the body heat.

You should also provide your birds with plenty of suitable shade for them to hide in. Yes, birds like to sunbathe, but they also need to be able to get out of the sun or they will overheat. You will also need to provide the birds with a constant supply of cool, clean water. On really hot days, chickens enjoy playing in mud puddles or in shallow pans of water. This helps keep their bodies cool. This is one of those instances where if you feel hot, you can bet that your birds are feeling even hotter.

Here in the South where we live, beyond the shaded areas, the mud puddles and extra water, we also provide our birds with fans. We buy the cheap, big, box fans and run an extension cord out to a shaded area and turn the fan on medium. It is not long before all the birds are lined up in the breeze. Another possible addition to your chicken pens would be misters. Misters are mainly used in the equine industry to help keep the horses cool during the summer months. These mister systems can be cut down to fit nicely in a chicken pen, or many of the big box stores now offer patio mister systems that can also be used to help keep your birds cooler. You can also freeze bottles of water and lay these about the pen. The chickens will literally cuddle up with them on hot days to help cool their bodies.

During this time of hot weather, we also cut back on the levels of fat and protein in their feed. Because most of our birds are kept for laying purposes, we generally do not drop below 16 percent protein levels, but this is usually sufficient to lean them out and also cut down on the extra heat produced by their bodies, so the birds stay much cooler.

Heat stress is a main cause of death in chickens during the summer months. If you see one of your birds not moving much, with her wings outspread and panting heavily, you know that she is getting overheated. Most of the time they are able to regulate their body temperature down through these actions, but it is possible for Miss Cluckers to get so overheated that she collapses from the heat. If you notice Miss Cluckers or any one of your birds in distress from the heat, pick the bird up and take her into the house. Run some cool water into the sink and slowly lower her into it up to her neck. Make sure you have a hand under her and over her to hold her wings down or you will get a bath right along with her. Once she is comfortable in the water, work the cool water up under her wings. They will usually come back around fairly quickly. After she has cooled, then you can take her out of the water and wrap her in a towel to start to dry off. Dry her gently with the towel, rubbing in the direction of feather growth. Then, place her in a cage to further dry on her own. After she has dried to some extent, release her back out into a shaded area to finish drying and so she can preen herself.

Dealing with heat and cold mainly comes down to common sense. Make sure the breed of birds that you have is suited for your climate. Make sure that the coop and pen areas are sufficient to protect your birds from weather extremes. Provide a constant supply of fresh, clean water and adjust their feed to meet the conditions. By doing these things, you and your birds will be much happier.

> You know you're addicted to chickens when . . . you will go out in the rain and get all wet just to make sure that none of your birds are getting wet.

Avoiding Predators

Predatory attacks on your flock are going to happen. You have chickens and there are many animals out there that like to eat them or their eggs. It is up to you to help keep them safe. Some predators are bigger than others and some are smarter or more determined. There are predators from the ground and from the sky. Each area has different predators that you may have to contend with and it can be very frustrating when you find that you have been raided in the night.

Your best defense is going to be strong, secure pens, but sometimes that is not even enough. The most common predators here in the United States are raccoons, opossums, skunks, foxes, dogs, snakes, hawks and bears. But there are also the two-legged kind, as in man—yes, man—and unfortunately he is one of the most difficult to catch. Each area deals with different predators and it is beneficial to talk with other local poultry raisers or your local Cooperative Extension Service to see what predators they deal with most.

A strong, secure pen is essential to the well-being of your flock. When building your pens, keep in mind that predators can come from every angle, and many of them can get through a really small hole. Your pen needs to have a ground barrier against digging or burrowing animals. Dig a trench about a foot deep and bury heavy, close mesh wire around the perimeter of the pen to keep animals such as dogs and rats from burrowing in. Your cage wire itself needs to be strong. Chicken wire is made for chicken pens but it is easily destroyed by larger animals. It is always best to place heavier wire, such as welded wire mesh or even chain link, over this. To help keep out snakes and smaller ground-based predators, put up half-inch square mesh around your pen to a height of about two feet. Because some predators come from the air, you should also put some type of netting over the top of your pens.

You might think that you are putting your chickens in Fort Knox, but to keep them safe, you just about have to. And if you free-range your poultry, then it is up to chance that you don't suffer from predatory attacks. How well you build your pens will be determined by what predators you have to deal with. There are many other options for keeping predators at bay, but nothing is fail-safe.

A good theory of pen building would consist of wire buried a foot deep around the perimeter, your main pen built with 2" × 4" welded wire mesh covered by one-inch mesh chicken wire, a nylon mesh netting over the top, an electric fence around the perimeter of your pen at six inches off the ground and another strand at three feet off the ground, a tightly built hen house or roosting house with a locking door and heavy wire mesh over all vent openings, with all your roosts away from exterior netting.

By building a pen this way, you can keep out most predators. Dogs, raccoons, opossums and skunks are kept out by the buried fence and the electric fence. Hawks are kept out by the overhead netting. Racoons and the two-legged predators are kept out by locked doors. Rats are kept out by the tightly built hen house and the mesh over the vent openings. And just because you live in an urban area doesn't mean that you wont be visited by a predator. Once they figure out that you have chickens, they will come.

What happens if you get raided in the night, or during the day, for that matter? How do you know what you are battling? There are some very common traits to predatory attacks.

- Dogs generally kill just for the fun of it. If you wake up or come home and find your birds dead and kind of piled in one area and not eaten, then this is usually from a dog.
- If you find your birds scattered about kind of leading off in one direction and sometimes partially eaten, you could probably bet on a fox.
- If you find a single bird killed during the day out in the open that is partially shredded, be looking to the sky.
- If you get up in the morning and find one of your birds lying by the fence with their head missing, count on a raccoon. If it happens during the day, then it may be the neighbors' cat.
- If you find that your prize-winning rooster has had his tail feathers messed up and only half their length is left, then suspect a rat getting into the roosting house.
- And if you wake up to find your pens destroyed, then you better call the game department because you probably have invited a bear to dinner.

Predatory attacks are going to happen, it is just a matter of when and how severe they will be. If you have an outside dog, this will help to deter some of the attackers, but you can also hang flashy items from strings over your pens to help keep out birds of prey. You can put up secondary barrier fencing to help keep out foxes and dogs. Use traps to keep up with the coons, possums and skunks. You just have to use your imagination to help outsmart these creatures. Unfortunately, sometimes it takes an attack to realize where the weak spots are in your pen design so that you can better secure your birds.

Gardening with Poultry

It is a well-known fact that chickens love to scratch in the dirt to find bugs and seeds. It is also known that they can decimate a flowerbed in minutes. So why not use these facts to your advantage?

In the world of chickens, their lives consist of four things: scratching, eating, pooping and squawking. If you are lucky they will also provide you with eggs and meat, but it is these first three things that we realize can be used for a common good. By putting your chickens to work doing what they love to do, you can reap additional benefits.

Chickens are systematic eaters. When turned loose on a plot, they will first go after bugs, seeds and grasses. They will then turn their attention to tender leaves and flowers followed by whatever else is left. If left to their own intentions, they can reduce a small plot to bare dirt in just a short time. But this is where human intervention comes into play. It is up to us, as the flock master, to place some controls over their behavior.

But you say, "When I let my chickens out to free range, they run all willy-nilly and destroy everything." That does not have to be the case. You have to employ flock management to reduce the amount of destruction caused by your birds. This can be accomplished in many different ways. Through the use of time allotments and portable cages you can effectively control what it is that your birds eat.

Let's say that you have an area that you would like to plant a garden in but it is just too overgrown with weeds and such. This is an easy one. By using a temporary or portable pen, you can allow the chickens to scratch to their hearts' content right down to bare dirt. Throw a bit of scratch grains out for them and let them go to work. They will reduce the area to bare dirt in no time and in the process add fertilizer by way of their droppings that can then be turned into the soil. Once the plot is to bare dirt, you can then turn the dirt over by way of shovel or rototiller. Then turn your birds back onto the plot so that they may scratch and peck at any weed seeds that are left, further breaking down the soil and adding more fertilizer. Many people employ this theory in their established gardens by letting the chickens onto the plot at the end of the growing season to reduce what is left to dirt and fertilizer. They will also plant cover crops such as rye or clover before the new growing season and allow their birds to reduce it to usable fertilizer before planting the next year's crops.

The big movement in gardening is to use raised beds so that you can easily reach across them from either side and so you don't have to bend over so far. When you are in the process of setting up your raised beds, if you keep in mind that you will want to use your poultry to your benefit, make the beds of a manageable size. Most raised beds are four feet wide and eight to ten feet long. If you build a movable pen that is of these same dimensions, it can be placed on top of the bed whenever you need your chickens to do their work. And if all your beds are the same size, then you could just move the pen from bed to bed as needed.

Another effective way to use your poultry around your garden takes a bit more planning but it is what we call "a moat." No, it is not the medieval water and gator-infested area that surrounds a castle, but it is the same principal. Instead of water, it will be dirt, and instead of gators, it will be chickens—but it will help protect your castle, the garden, from marauding bands of enemies: bugs. Around your garden area, install a four-foot high fence. Approximately six feet out from this, install a second fence and place screen over the top between the two fences. You can allow your chickens to freely roam this "moat" area all day long. Any roving bugs that try to enter your garden will quickly become consumed and limit invasion into your garden area. It will also keep the perimeter area clean and tidy, as any weeds that try to grow will also be consumed.

OK, so let's say that you do not have a garden where you can put your poultry to work. What else can they do for me? Do you have shrubbery beds around your property? Do you hate getting in there and trying to weed them out? Let your birds do it for you. Again, your chickens are going to go after the weeds and bugs before they go after your bushes. If you limit their time and exposure to a certain area, then they will accomplish what it is you set them out to do.

But as chickens are quite clumsy, I would not turn them loose in your blue-ribbon orchid patch. You do have to use some discretion when turning them out. If you are looking for some general maintenance-type weeding, turn your birds out an hour or so before they go to roost in the evening. This will limit their exposure time. Do this a couple of times a week and you will save yourself hours of picking weeds. If there are areas of your yard where you do not want them to go, put up a simple temporary fence to keep them out.

Chickens really are master gardeners. They clean an area to bare dirt. They fertilize as they work. They loosen the soil. They can make things grow. They can weed and they are happy about what they do. What more could you ask for? And as they toil away the hours, you can sit and watch them with a cold drink in your hand and enjoy their antics.

Starting a Chicken Farm

A man from the city wanted to start a chicken farm of his very own. He was sent to a local hatchery to purchase some chicks to start the enterprise. He went in and purchased 500 fine White Rock chicks to start the farm. The next week he came back to the hatchery and purchased 500 top quality Rhode Island Red chicks. The following week it was 500 Cornish chicks and so on and so on for many weeks.

Hatchery man: Wow, you must be starting a huge poultry farm with all of these chicks!
City slicker: Not so big really. I'm just having a little trouble with this first crop. I can't tell if I'm planting them too deep or too close together.

Composting with Chickens

You dutifully feed and water your chickens each day and provide them with secure, dry shelter, and in return they provide you with eggs and meat. But did you know that you can put your chickens to work? It is true, and one of those ways is by letting them tend to your compost heap.

Chickens are great at turning over dirt and debris, happily clucking away as they scratch for bugs and other yummy morsels buried within. And what goes in must come out, which helps further fertilize the pile. Their constant pecking, scratching and pooping quickly breaks down all types of organic matter, which in turn breaks down faster into usable compost for your garden and flowerbeds.

There are many ways to accomplish this process. You can build fancy compost bins, but we do it in a very rudimentary fashion, which works well for us. Within our main community pen area, all of our birds are kept on dirt except the really young chicks, which are kept in adjoining grow-out pens on pine shavings for the first six weeks. Our roosting house is a separate structure from the hen house and is also dirt floored. Once a week, the roosting house is raked out and into a pile in the middle of the common area. To this is added leaves and grass clippings from the yard and flowerbeds. Our birds are also given all of our household leftovers. The chickens all love to scratch and dig through this pile, and in the process they mix the layers together and break down the materials. Every couple of days, I spend a bit of time, as the birds are getting ready for bed, raking this all back into a nice pile. It is great exercise for me and accomplishes the turning of the compost pile. During the warm, dry days of the year, I will wet the pile down so it can sit wet overnight before the chickens tear back into it in the morning. Every couple of weeks, I will rake the shavings out of the chick pens and add those to the pile also. This goes on for two to three months, until a nice pile is formed. I then start a second pile. Meanwhile, the chickens continue to scratch and turn the first pile and I continue to rake it back into a pile and make sure it is good and wet. After another month or so, I am able to take this first pile out to the garden and flowerbeds as nice fertile compost.

Just about everything goes into the chicken pen prior to being disposed of in other ways if necessary. Branches, sticks, logs, food products, yard waste and more all goes to the chickens for picking over. As I rake it all into a pile, I sort out the bigger stuff, which goes over the fence into the long-term compost heap for decomposing. We watch closely what goes in the pen so the chickens aren't given anything harmful. But as we always say, chickens are pigs with feathers.

You could certainly expand on this idea by building open compost bins that the chickens can get into to scratch around. This would save you some time and energy from having to rake the compost back into a pile every couple of days. Our birds are also wormed every six months in case they pick something up from the soil or compost. We allow our birds the luxury of meat products, especially in leftovers, but we do not feed them raw meat; also, this type of food is not directly added to the pile. Nothing is added that can mold or otherwise has the potential to cause possible toxins being added into the pile.

This all leads to happier chickens because they think they are getting something special. It makes us happy because we get something else out of our birds and it provides us with nice compost in the end.

3 GETTING STARTED WITH CHICKS

All About Chicks

What came first, the chicken or the egg? That question has been asked for hundreds, if not thousands of years. The answer is, whatever route you choose to take.

Most people, when starting out wanting to raise chickens, will pore over the countless hatchery catalogs, looking at all the available breeds, and then pick one or two and order their day-old chicks. We get many e-mails from people across the globe that have done just that; they ordered their chicks, the chicks arrive and then they are like "Now what do we do?" There is a little bit more to raising chickens than just ordering them from a catalog because they look good.

There are many ways to obtain newly hatched chicks. Obviously, you can order them from a hatchery, but come the first days of spring, signs will pop up on the road side and ads will be placed in the newspapers offering chicks for sale. The local farm and flea markets will be flooded with chicks. You may even know a fellow chicken breeder who is hatching chicks as you are reading this, and if you are really adventurous you could try your hand at incubating eggs (more on that later in this chapter). But where is the best place to get them? What is the right thing to do? It can all be so confusing.

Many people who are just starting out will have their mind set on a specific breed because that is what they remember Grandma having on the farm when they were kids. After all, if Grandma had them then, they must have been good. But chickens have come a long way since Grandma's time. They have been bred and cross-bred to bring out specific traits that are more desirable and bred to take out the traits that some might find less desirable. New breeds have been introduced and many of the old heritage breeds have become almost non-existent. There are chickens out there that have been so extensively bred that they do not much resemble a chicken when it comes to doing what chickens do naturally.

So, before you go off and run to the local feed store or order one hundred chicks from a hatchery of a breed that Grandma had or just because the picture looks good, STOP! Now take a deep breath. Put the catalog down and step away. There are a few questions that you need to first ask yourself. Actually, there are a lot of questions, but we will take them one by one the best we can.

Let's start off with the basics. First, what is it ultimately that you want out of your flock? There has to be some type of goal in mind. If you are looking for a few birds to produce eggs for your family for Sunday morning breakfast, then the options are wide open. But, if you want a nice flock of birds that are going to produce enough eggs so that you could sell them to friends and neighbors, then your choices diminish. If you want a bird strictly for meat, there are only a few choices. Maybe you are looking for just a few birds but the grandkids come over on weekends and they like to go visit with the chickens; you will then want a breed that is calm and easily handled. What are you going to do when these birds get older and slow down on laying? Do you want to breed your own chicks or are you going to purchase chicks each time you need to replenish your flock? Maybe you want to hatch out a few chicks that you can sell each spring. You will need a breed that will either go broody easily or buy an incubator to hatch your own. It is best to write out all your desires for your birds and then peruse the catalogs to see what breeds fit best to your needs.

Another great resource for checking out the attributes of different breeds is Henderson's Handy Dandy Chicken Chart (www.sagehenfarmlodi.com/chooks/chooks.html).

Next is space. How much covered area and open area are you willing to relinquish to the poultry flock? If you live in an area that experiences hard winter weather, you will need sufficient covered area for the birds to roam around in during the day. If you live in the hot, humid South, you will need an area that provides plenty of shade. Are you planning on free-ranging your birds? Just a few chickens can quickly turn your beautiful flowerbeds to bare dirt.

These are things that need to be considered prior to picking out a batch of those cute little fuzz balls. Still, there is sure to be a breed that fulfills most, if not all, of your desires, but do your research first. Ask other poultry growers what their experiences are. Visit a few local farms and see what breeds they have and what the temperament of the breeds are. Maybe even visit a local poultry show or your state fair. Once you have decided on a breed, the question becomes whether to buy hatchery chicks or locally-hatched chicks. There are advantages and disadvantages to both.

Hatcheries have the ability to send you as many chicks as you need, in the breed you desire, and usually sexed with a 90+ percent accuracy. Some of the disadvantages to hatchery chicks is that they usually have a twenty-five chick minimum, and they are shipped through the postal service, which can delay or lose your chicks or treat them roughly, resulting in a high number of deaths or damaged birds. Your local feed store may be a viable option for hatchery chicks and you would be able to buy them in a smaller quantity and they are usually sexed. Choosing locally-hatched chicks generally allows you to get your chicks quicker than through a hatchery and without the risk of loss or damage as well as in smaller quantities if needed. At the same time, you never know what you are really getting unless you know the breeder well. Just because a farmer says he has Rhode Island Reds doesn't necessarily mean that they are true Rhode Island Reds; they may be Red Sex-Links or a mix breed that resembles a Rhode Island Red.

Furthermore, many backyard poultry people do not offer sexed chicks unless they are of sex-linked or auto-sexing varieties. Hatcheries generally have healthier chicks, but healthy chicks are also available from a reputable local breeder. Look for someone who is National Poultry Improvement Plan (NPIP) certified. This will tell you that they have their breeder birds checked once a year for serious diseases and ask to see their breeder stock and facility. A breeder may or may not allow you to see their facility, depending on the level of bio-security they have in place. If they are willing to show you their birds and facility, look around. Do the birds

Visit a few local farms, a poultry show, or state fair and see what breeds they have and what the temperment of the breeds are.

appear healthy? Are the pens relatively clean? Does the breeder stock look like the breed you are seeking?

Another disadvantage of hatchery birds is if you are looking for a bird that you might be able to show down the road. Most poultry judges really frown on hatchery birds and will usually quickly disqualify them. They usually will not meet the standards set forth by the American Poultry Association (APA). That is not to say that you couldn't take a hatchery bird and selectively breed it and eventually show some of the offspring.

Hatchery birds have a tendency to be heavily bred to resist going broody. If a hatchery has parent stock that tends to go broody, then they have hens that are out of egg laying for a number of weeks and not producing eggs for incubation. If you are looking to perpetuate your flock, you will probably be better off with home-grown birds unless you plan to incubate the eggs yourself.

Did you know?
A fertile egg begins to grow into a chick once the temperature of the egg reaches 88°F (31°C).

Setting Up Your Brooder

So, you have finally decided on a breed that best fits your needs, you have made the decision of where to buy your chicks, now what? What are you going to do with one hundred chicks? Or fifty? Or even ten? Your new chicks are going to need special care and housing for the first few weeks.

If you are getting day-old chicks, either from a hatchery or from a private person, then you will need a brooder set up prior to their arrival. A brooder can be as simple as a cardboard box with a light over it or as complex as you wish to make it. The main requirements are that the chicks have adequate heat, space, feed and water. Your heat source can be a regular light bulb suspended over a box. The temperature at about two inches off the bottom needs to be about 95ºF (35ºC). You will want to drop the temperature about five degrees Fahrenheit each week until you reach ambient outdoor temperature. This can be achieved by simply raising the lamp higher each week. Here in the South, we find that an 85-watt floodlight works just fine for our chicks. It provides enough heat to keep them comfortable. In the colder climates, you might find that you need a 125-watt or even a 250-watt heat lamp bulb to provide enough heat. There are also heaters made specifically for brooding chicks, so whichever is your choice to use, watch your chicks for the right temperature.

Now, we use red colored bulbs and you are probably asking why. You can certainly use clear bulbs, but there are some advantages to using red. The white light of a heat lamp bulb is very hard on the chicks' developing eyes, but with red bulbs it is not so bright in the brooder area. If you are using clear bulbs and you notice that chicks are having their feathers picked out by other chicks then you will want to change to a red bulb to help control this urge.

Another very important thing to remember with any heat source is it has the ability to set things on fire. I don't know how many times I have read where someone's barn or chicken coup burnt down because their brooder heat source was too close to the litter or it fell down. A heat lamp throws off a tremendous amount of heat, hence the name. You must use a lamp shroud that is rated for the wattage of bulb that you are going to use. Most commercially-produced brooder lamp shrouds come with a wire bail on top. This is there for a reason. Hook a rope or small chain through this bail and hang the shroud by this. I also run the cord up overhead and attach it as a safety line in case the other line fails. Your brooder heat source should never get close enough to the litter to be able to start a fire. Yes, accidents happen, but in this case they do not have to.

Some people use a heating pad on the floor of the brooder. This is not a wise idea. The chicks have no way to get away from the heat if they get too hot. I have heard of way too many cooked chicks from people using heating pads. Do yourself and your chicks a favor and put the heating pad back into the cupboard.

There is no need for a thermometer, as the chicks will tell you if the temperature is right. If the chicks all huddle together, then they are too cold. If the chicks are all pressed up against the walls, then they are too hot. If they spread out and sleep all over the place, they are just right.

Space is very important. Chicks will grow fast and can quickly outgrow a box, but if you start out too big, then you will not be able to provide enough heat to keep them warm easily. The general rule of thumb is a half a square foot of space per chick for the first two weeks of growth.

We have found that we can comfortably brood twenty-five chicks in a clear plastic tote that is 105-quart size, which measures roughly seventeen inches wide by twenty-seven inches long and thirteen inches deep. We brood our chicks indoors, so we use an 85-watt red floodlight for the heat source. This works for the first two weeks before they are moved to an outdoor pen for further brooding.

Food and water need to be provided constantly to the chicks. In the plastic totes that we use, we're able to place a one-quart feeder and a one-quart waterer. Newborn chicks will eat chick starter crumble. You can feed them medicated or non-medicated feed, depending on your preference. Medicated feed contains a coccidiostat. This helps combat coccidiosis, one of the most common diseases to affect chicks. It's imperative that the chicks always have fresh, clean water. You may find that you need to change the water three or four times a day in order to keep it clean.

With day-old chicks, it is advisable to place marbles or small stones in the dish of the waterer to help keep the chicks from drowning. These can be removed after four or five days.

Another thing to be aware of with day-old chicks is that their leg bones are very soft. For this reason it is necessary to place a non-slippery surface on the bottom of the brooder. It is best to use paper towels for at least the first three days. This gives them a surface that has traction and is also a surface that is easy for them to see feed on. After three or four days, when the chicks are all eating from the feeder, they can then be transitioned onto pine shavings or another suitable litter. Do not ever use cedar shavings around your birds as cedar gives off oils that can be toxic to chickens.

You know you're addicted to chickens when . . . you come home from work and your wife is wearing a chicken suit so you will spend time with her!

The Life Stages of Chickens

Let's take a look at the various growth cycles of the chickens from day-old up through young adults to give you some idea of what to expect and how you can prepare for the joy that lies ahead.

0 TO 6 WEEKS OF AGE

That magical day arrives when the post office calls you to tell you that your chicks are in and to come and get them before all the chirping drives them all postal. You hurriedly drive down to the post office and they gladly present you with a noisy box and try to hurry you on your way—but don't rush off quite so fast. Open the box there in front of a postal worker to make sure of a safe delivery. Always expect to see a couple of dead chicks. The hatcheries add extras that we call "packing peanuts" for additional warmth and also to cover any possible losses from transit. What a wonderful sight to see all those fuzzy little chicks. Chicks do what is called imprinting, which is to say that the first thing they see becomes their mama. You are now their mama and they will rely on you for their needs.

You rush home to your waiting brooder. As you remove each chick from the shipping box, be sure to dip its beak into the water so that it knows where to find it. Let your chicks settle in for a couple of hours. They will chirp wildly for the first hour or so, which is natural until they get used to their new surroundings. Remember, these poor chicks were hatched, sorted and stuffed into a box all within the first hours of life. Then they were placed on a bumpy truck for an endless ride to your post office. You peeked at them and then covered them back up and then they were off on another ride to your home. They are taken from their box, dunked in water and then turned loose in a foreign land. You would be chirping too!

Chicks will sleep a lot for the first week or so of life. They use up energy fast and do not eat a lot those first few days. As they begin to eat more and more, they will also be up running around more. Your main concerns at this point are that they have plenty of feed, clean, fresh water and adequate heat. The other very important thing is to make sure their paper towels or litter is clean and dry.

Water is a chick's best friend and also its worst enemy at this stage of life. They need clean water to drink but they also need dry litter. If their litter gets wet, it can cause all sorts of problems from chills to respiratory issues to coccidiosis, so it is imperative that their litter remains dry.

At this stage, we always find ourselves sitting and watching the chicks, which can turn into hours of enjoyment. Don't be afraid to interact with your new chicks. Place your hand into their brooder area to their height. They will run away in fear and all clutter together. Leave your hand there and soon a lead chick will come to investigate. As it gets close to your hand, move your fingers slowly and rub its chest.

They soon discover that this is a good thing and will come to your hand each time you reach into their area. When they get used to your hand being in there, you can then start picking them up to play with them. It is not long before they almost demand to be picked up and held.

While you are watching your chicks, pay close attention to their little butts. Make sure that they do not have poop pasted over their vents. This is called "pasty butt." If they do, you will need to remove it, otherwise they can get plugged up. If it is really dried on, you can take a warm, wet washcloth and moisten the poop, and then with your fingers gently break it away from their skin. Be careful so that you do not make them bleed, as this is a very tender area. If the poop is moist, you can usually wipe it away with a tissue.

Also watch your chicks over the first twenty-four hours for crooked toes and splayed legs. Both of these can usually be easily treated if caught right away and both will be covered in chapter five.

Most of all, be on the lookout for disease. Coccidiosis is going to be one of your main concerns. It is contagious and can wipe out all your chicks in a matter of days if not treated. Damp or humid conditions make your chicks more prone to getting coccidiosis; that is why it is so important to keep their litter clean and dry. This is covered in more detail in the disease section of chapter five, but watch for chicks with ruffled up feathers, lethargy, no interest in eating, and blood in their poop. These are signs of trouble, but if caught early enough, coccidiosis is readily cured with medications in their water. Just because you feed your chicks medicated feed doesn't mean that they can't get coccidiosis; it just means that they are less likely to.

Respiratory issues are the other main disease factor that you must watch for. Your chicks will sleep face down on the litter. If the litter is not clean, there will be higher levels of ammonia, which can burn their delicate lungs. If the litter is damp, the amount of ammonia given off by the droppings is significantly more. Watch for your chickens to be gasping for air. It will be like they are slowly panting. This will tell you that either they are way too hot or the ammonia levels are way too high in the brooder and causing respiratory problems.

As touched on before, your chicks will tell you if they are too hot or too cold. If they are all huddled together, then they are too cold. If they are all as far from the heat source as possible, they are too hot. They should move about freely and sleep all spread out over the floor of the brooder.

For the first week of life, chicks need a temperature of 95°F (35°C) degrees. Every subsequent week, the temperature needs to drop by 5 degrees Fahrenheit until you reach the ambient outside air temperature.

Your chicks will continue to grow and soon they will be developing their feathers. They will also discover that they have wings. By this time they are about two weeks old and will probably need to be moved to a bigger brooder area. You should have the temperature in the brooder down to around 85°F (29°C). Chicks will fully feather out in about six weeks, at which time they can be moved to a grow-out pen. Many people will continue to have heat on their chicks at this age. You are not doing your chicks any favors by providing heat longer than necessary. Chicks and chickens produce body feathers relative to the air temperature. If you keep your chicks all nice and toasty warm and it is only 40°F (4°C) outside, they will not develop enough down feathers to keep them warm once you move them to the grow-out pen.

Chicks sleep a lot during their first week of life.

It's actually better for them, the faster you can get them off the heat. Think of a mother hen. When she has new chicks, she will be running around with them by the third day after hatching. If the chicks get cold, she will set down and let them climb under her until they are warmed up, and then it is back to pecking and scratching. If you take naturally-hatched and brooded chicks and place them against artificially-hatched and brooded chicks that were kept under typical brooder conditions, you would find that the natural chicks are further along feather-wise than the artificially-hatched chicks. This is because they are adapting to their surroundings. If it is 85°F (29°C), then they will have feathering to cope with warm temperatures. If it is 40°F (4°C), then they will have feathering to cope with colder temperatures.

By eight weeks of age, the chicks should be pretty settled into a routine of where to sleep at night and where to play during the day.

The chicks that we hatch artificially are kept in a brooder box inside the house with an 85-watt floodlight bulb over them for the first two weeks. After this point, they go outside into a small grow out pen with an 85-watt bulb over them. By the fourth week, the lamp gets shut off in the morning and does not get turned back on until nighttime. By five weeks, our chicks are usually fully feathered and sleeping on roosts with no light at all. And this is during our winter months where the temperature actually does get down into the teens at night. If by chance we hatch chicks during the summer months, the lighting program is stepped up to where they only have a dim light source and no heat source by the time they are three weeks old and by four weeks old they are trying to roost like the big chickens.

Q: What is a chick after she's six days old?

A: Seven days old!

---

6 TO 14 WEEKS OF AGE

You have your chicks and you have been growing and nurturing them through their first few weeks of life. Those cute little fuzz balls are starting to feather out, learning to fly and finding themselves within their little flock. You are finding that their brooder area needs to be expanded almost weekly and they are spending more time away from their heat source as they explore and learn.

Continue to reduce the amount of available heat as they age by raising the heat lamp more and more. Let the chicks tell you if they are too cold or not. By week six, at the latest, you should have no heat on them at all unless you have a really cold night, then you could put the light back on them through the night and then remove it in the morning.

If your days are nice, by the age of six weeks you can start to let the little chicks venture out of their brooder area and into a protected run for additional exercise and to get used to the sights and sounds of the outside world. Make sure that the run area is secure and covered with at least wire screen, or only let them out when you are present, as these little ones are easy pickings for predators.

Being let out of their brooder area for the first time is quite overwhelming for them and they may be very hesitant to venture out. You can scoot them out to show them that it is OK. Once outside they will soon take to dust bathing and sun bathing in between running and flittering about.

By this time you should be able to start to tell which ones are boys and which ones are girls, if you started with mixed-sex chicks. If you started with all pullets, then you hope that all are in fact girls.

By eight weeks of age, the chicks should be pretty settled into a routine of where to sleep at night and where to play during the day. Their brooder area should now be a roosting area, whether you use the same pen or move them to a grow-out pen area. You should have various roosts set up for them that are about eighteen inches to two feet off the ground. Don't be surprised if the chicks play on the roosts during the day but still want to sleep on the ground at night, as this is normal behavior. I find that if I go in to their pen at night and place the youngsters up on the roosts, this encourages them to roost on their own at night.

At eight weeks you can start giving the chicks a few treats, but only in moderation. They are still growing very rapidly and so you do not want to replace their normal diet with a bunch of nutritionally deficient foods. A few slices of crumbled bread or a bit of scratch feed every few days would be more than enough. Use these times of giving treats to get the chicks to come to you and maybe even eat out of your hand. The more you get them used to your presence, the easier it will be in the long run when working around them.

This stage of the chick's life, from six weeks to about fourteen weeks, is mainly relegated to growth. Keep them on a good quality chick starter/grower or flock grower feed. Watch for any abnormalities in growth such as runting, or feathers not growing in the way they should. Also watch for nutritional deficiencies to pop up that might show, such as crossed beaks, crooked legs or lack of weight gain.

For production type breeds, at eleven weeks, you are about half way to seeing the first fruits of your labor. With meat-type breeds, at fourteen weeks you should be getting pretty close to sending your birds to what we refer to as Freezer Camp. For bantams and ornamental breeds, growth rates and maturity happens at varying stages, so use this time to get to know your specific breed.

14 TO 22 WEEKS OF AGE

I am of the belief that every person of childbearing age, male and female alike, should raise a mixed-sex flock of chickens to adulthood before they consider having children of their own. It is amazing how much those cute little fuzzy chicks can mimic the lives of humans as they grow, especially during what I like to refer to as the teenager stage, that magical age somewhere between fourteen weeks and twenty-two weeks.

You have worked diligently with your chicks from day one, doing your best to keep them healthy, sheltered and fed. You have built a strong bond with them and they know your voice and look forward to your visits. You have handled them and nurtured them along the way. You are proud of that little batch of chickens that have learned so well. Yes, there has been a squabble or two along the way, but disputes were easily solved. Grandma comes to visit and remarks on how well you have done raising your little boys and girls, and then one day it happens . . .

You wake to this shrill noise resembling that of a dog squeak toy. It repeats over and over. It catches you off guard and you are thinking, "What in the name of Hannah is that?" You race to the pen to find one of your cute little chickens standing on the highest perch, puffing up its chest and craning its neck, only to let out this pitiful, drawn-out squeak. It has begun; your birds have reached puberty.

Within the coming days you will realize that all your hard work has gone out the window. It is like they have lost their mind. The boys are clumsy, gangly and kind of dorky looking. They run around chest bumping each other and trying to look cool for the girls. The boys and girls alike are starting to develop their adult plumage, the boys get their tails and the girls get their fluffy butts. The girls will try to separate from the boys and stay in their own little cliques.

The days wear on and the hormones kick in. The boys are trying to establish themselves within the flock. If you have a mixed-age flock such as mine, the boys will try to get near the older ladies and ladies will quickly put them in their place with a sharp peck and many times a scolding. Testosterone has taken hold and the boys want to show their manliness. They are clumsy and unskilled at mating and grab the girls by the head or by a wing and drag them around. They may just come away with a beak full of feathers from the failed attempt. The younger girls squawk and run like those boys are the worst things to ever come along.

Have no fear, folks; just as with our own children, this stage only lasts for a short while. In human children it goes from about age thirteen until they are about forty. In chickens this stage only lasts about three weeks, thank goodness.

In a mixed-age flock, the transition from boy to man and girl to woman seems to be much easier on all involved. The boys and girls have role models. An older rooster will show the boys what it means to court and woo a lady so that she is ready for mating. The ladies will teach the young boys how ladies are to be treated if they want to have any chance of mating with them. If the boys get out of line and start fighting a little too rough, then either a good rooster or a dominant hen will break them up and send them on their way.

With my breeding birds, if I decide to change the rooster or bring up a new flock of girls, I always try to add young to old either way. I will put a young rooster with older girls or I will put younger girls with an older rooster. I feel this helps create a positive learning atmosphere for all. The older girls will train the young boy how they want to be treated or an older rooster knows how to treat the younger girls through past experience. If you are dealing with a single-age flock, the process takes a bit longer but they all soon learn and once again there will be peace in the valley.

If you choose to only have a flock of girls, then much of this becomes a moot point. Yes, there will be pecking orders established and you will have the dominant girl and the submissive girls, but for the most part girls get along quite well within a single-sex flock. If you are raising a flock of meat birds where the majority, if not all, of them are boys, then the puberty stage seems to be delayed, but hopefully by then they are all resting comfortably in your freezer.

Chicks can mimic the lives of humans as they grow, especially between 14-22 weeks.

After about three weeks, all this nonsense should settle down. The boys will find their place within the flock; the girls will start mingling more. The boys will be perfecting their craft and smoothness with the girls and there will be much less squawking and carrying on. The boys' voices will deepen and their crow will become more like that of a chicken instead of a squeak toy. The girls' butts will get fluffier and widen as they get closer to starting to lay. If you have an older, more experienced rooster in your flock, he will start paying more attention to the young girls, and about two weeks before they lay their first egg he will start trying to mate with them and claim them for himself.

In most layer breeds of chickens, they reach this sexually mature stage at about twenty-two weeks of age. In the production breeds such as the Sex-Links, they may reach maturity at around seventeen weeks. But you may find that in some standard breeds such as Orpingtons or Brahmas, and in the ornamental breeds, that maturity may not come about until they are around thirty weeks. All this nonsense of adolescence, puberty and maturity will vary from breed to breed, but just know there is light at the end of the tunnel and they will not always be this way. Their loss of reasoning is only short lived.

YOUNG ADULTS

By now you have earned a reprieve. You spent countless hours watching over your little fluff balls, making sure that everything was OK. You may have even found yourself checking on them in the middle of the night just because you thought maybe something wasn't right. You cautiously weaned them off of their light and heat wondering every moment if maybe they were too cold or even afraid of the dark. You watched them grow by leaps and bounds as they went through that gangly awkward stage. You bit your tongue and showed your frustrations as those cute little fuzz balls turned into deaf teenagers, running willy-nilly and finding themselves within the flock. You listened to countless hours of squeaking and squawking as the boys gained their voice and the girls ran from them. Then everything kind of seemed to change over night.

The boys are off on their own and trying to gain themselves a group of girls. The girls are showing more attention and affection towards the boys. The crows are now stout and loud and the girls cackle and coo. And then that magical day arrives. You have seen the girls playing in the hay in the boxes, scattering it here and there and wondered if they may be planning for their future. That morning you hear a cackling like never before. The boys are raising a ruckus as well. You rush out to see what on Earth could be the matter. No predator from land or air. Everything looks fine from what you can see, but then there in the nest box, that one on the bottom tucked in the corner, you spot it—a small brown orb. You rush in and with the gentlest of care cradle your first egg. A smile washes over your face and you rush back to the house with that first egg firmly in your grasp. You hurriedly call your mother, your brothers and sisters, your relatives and friends to announce that you have graciously received your first egg—that all that work over the last many weeks has finally paid off. Walking back out to the pen you congratulate your flock and give them a special treat for giving you such a wonderful surprise.

Yes, your flock has finally matured. Within a week or so the rest of the girls should start laying their first eggs also. These first eggs are generally smaller in size and sometimes misshapen as their reproductive systems get into egg-making mode. Once the girls have been laying for about three months, you should be getting full-size eggs for the breed you have chosen. You may also find that the color of the egg will change. For brown egg layers, this is usually a transition from a light shade to a darker shade as they age.

On both the boys and girls, just prior to maturity, their combs will also darken and become larger. On the boys you will see small nubbins of their spurs starting to grow. On the girls, their butts will get fluffier and wider.

If you are like most people, you got your chicks in the spring somewhere around March or April, which means by now it is around August. In the northern states, fall is in the air; in the southern states, we just wish fall would hurry up and get here. But soon the days will be getting shorter and your sweet girls that just started laying

When the chicks reach young adulthood, the flock will be mature, and hens begin to lay eggs.

will go into their first molt. As the dog days of summer wear on and the nights are getting longer, it is very possible for your girls to stop laying. I know, they just started . . . so how can they stop now? That is nature's way.

The boys and girls will both be putting on more weight and their feathers will fill in more and more. The girls' laying will remain sporadic unless you have a hybrid breed, which could very well continue to lay right on through winter. This first fall and winter will be a time of growth both externally for the boys and internally as well for the girls. Come springtime, your flock will be nearly fully matured. The girls will start back in laying around February and be entering into their first full year of production.

It is a long process but worth it in so many ways. You can look at your fully-grown flock and be proud of what you have accomplished. And when the flowers of spring start showing themselves, do not be surprised if one of your hens comes crawling out of hiding one day with a little flock of her own. You will know that the circle of life has been completed and you have played a major part in this wonderful happening.

Did you know?
It takes about four pounds of feed for an average hen to lay a dozen eggs.

Broody Hens

Much discussion comes up about broody hens, how and why it happens and what is our role in the cycle of life. Before the advent of the incubator, it was left up to the hens to perpetuate the flock. If you had a good broody hen, she was about the most important bird in the flock. She is the one that would sacrifice her time to set on a clutch of eggs and hatch them out. She would care for the chicks until they were old enough to be on their own. With the first practical incubator coming onto the scene around 1879, it changed the world of poultry forever. Chicks could now be hatched artificially and the role of the hen was to be relegated to the job of laying eggs. Through selective breeding, most modern breeds of chickens have had the instinct to go broody bred out of them.

A hen cannot be made to go broody! A hen, if she is going to go broody, will generally do so in the early springtime. This is the natural cycle of life. In the poultry world, a hen will molt during the fall or early winter, at which time she loses most of her feathers. Her new feathers will come in thicker and healthier to get her through the winter. During her molt, she will either stop laying or extremely slow down the egg laying. Once her molt is finished, usually after thirty to forty-five days, she will begin to lay again. But she is also entering the winter season in which the daylight hours are shorter, so she will generally lay only sporadically. This is her time of rest. But after the winter solstice (December 21), as the days begin to slowly get longer, her body's natural time clock kicks in, and at the first signs of spring, the natural instinct is to go broody and hatch a clutch of chicks. This is a natural instinct, not a learned action.

As people are now starting to get away from hatchery birds in favor of the more traditional heritage breeds, we are seeing a greater occurrence of naturally-hatched chicks. For the small backyard poultry enthusiast, a broody hen can be invaluable.

But how do I know if a hen is going broody? A good broody hen will pick a nest area that is generally the most secluded. In a hen house, this might be a far bottom corner nest or the highest nest she can comfortably get into. For free-range birds, a hen might pick an area under a building or up in the corner of the hayloft. A hen that is going broody will remain on the nest for longer periods of time each day. If you try to move her, she will likely growl, fluff up her feathers and peck at you. She will be driven to sit on that nest whether there are eggs in it or not.

A broody hen will lay a clutch of eggs. This could be from one egg up to twelve or more. These will be the eggs that she will sit on until they hatch. If other hens lay their eggs in her nest, she will also take those eggs for hatching. A typical hen will generally sit on ten to twelve eggs proportionate to her body size. You will not be able to have a little bantam hen sit on ten large fowl eggs. Once a hen has laid her clutch of eggs, she will stop laying eggs and begin her set. She will then sit on her eggs for twenty-one days until the chicks hatch. This is confusing for most people, when they realize their hen has gone broody and then after twenty-one days they wonder why the chicks have not hatched. The actual incubation process does not really get into full swing until the day after the hen lays her last egg, so it is very difficult to determine the exact day that the chicks should hatch.

During the days that the hen is laying her clutch of eggs, she may remain off the nest for great lengths of time. She will return to the nest at various times to roll the eggs. As time gets closer for her to start her set, she will spend more time on the nest controlling the growth rate of the embryos. She does this by moving the eggs closer to her belly or farther away. Realize that she has laid her clutch of eggs over a number of days but her goal is to have all the chicks hatch within a twenty-four-hour period. So the first egg she laid may be twelve days older than the last egg she laid. Through careful manipulation of the eggs, she can speed up or slow down the rate of growth of the chick.

During her set, the hen will constantly talk to her growing embryos. She will roll the eggs up to fifty times per day to exercise the embryo and to prevent it from sticking in the shell. She will also control her body's humidity by spreading out more or tucking the eggs closer under her.

A setting hen will generally only leave the nest once a day to eat, drink and poop. This will usually only take her about half an hour and then she will return to the nest. If a hen has to travel a great distance from the nest to get food and water, she may only leave the nest once every few days, which is not healthy for the hen. If you have a hen that has entered her set in a remote location, you should provide her with her own separate food and water nearby.

A broody hen will be fiercely territorial. She will growl at and attack just about anything that comes near. It is her natural instinct to protect those chicks at just about any cost. I have had many people write and ask why a hen would eat her unborn chicks? This is part of the survival instinct. She will eat her chicks if she feels there is a serious threat or if the embryo stops growing for some reason. Most hens will kick eggs out of the nest that have not developed or have stopped developing, but others will eat the egg. If you notice that an egg is missing from the nest, suspect that the hen has eaten it. If your hen is in a remote location, if at all possible, it is best if you can install some sort of temporary cage around her to help protect her from predators. If she is in your hen house, if at all possible, move her to a separate area that would help protect her and the chicks from the other birds. We'll cover more on moving the hen and her nest later in this chapter.

Your broody hen will sit on her eggs for approximately twenty-one days. During this time she will roll the eggs and move them into different positions under her. She will regulate the temperature and growth rate of each embryo and she will constantly talk to the developing chicks. Once the eggs begin to hatch, she will stay on the nest for up to three days. After three days, the chicks that have hatched will need to eat and drink, so she will leave the nest.

Any eggs that have not hatched by this point will be basically abandoned. You can take the remaining eggs and discard them along with the shells of the hatched chicks. Check the eggs that did not hatch for signs of life. Hold each egg up to your ear and gently tap on the shell with your fingernail. If you do not hear any scratching or chirping from inside the shell, it is pretty safe to throw out the egg. If you do hear sound from inside, then you have the choice of leaving the egg in the nest and hope that the hen will return to the nest enough to keep the egg warm until it hatches, or you can take it and artificially incubate it. If you hear chirping inside the egg, you can be pretty sure that it will hatch within a day. It can be placed in a very rudimentary incubator such as a small box with a lamp over it. Use a thermometer to measure the temperature, which should be right at 99.5ºF (37.5ºC). If the chick should hatch, then it can be placed back with the mother hen once she goes to nest for the night.

Once all the chicks have hatched, it is up to the mother hen to teach the chicks those things that are not naturally bred into them. She will teach them how to find food and water by scratching in the dirt. She will teach them to come when they are called. And most importantly, she will teach them about predators, both from the ground and from the sky. The hen will stay with the chicks for about six weeks total. At about five weeks, she will start distancing herself from them more and more. By six weeks, the chicks should be pretty much on their own. The chicks will continue to want to hang around the hen, but she will peck at them to make them go on. About this time, your hen should start laying eggs again and rejoin the flock in a normal capacity.

SURROGACY

A good broody hen can be an invaluable addition to any farm flock. Not only can she hatch her own chicks, but a good broody hen can, and will, also hatch chicks from other hens and even from other breeds of fowl.

A hen should be able to easily accommodate up to a dozen eggs relative to her body size. If she is a bantam breed, she should be able to handle a dozen bantam eggs, but a good broody will take on the task of hatching other eggs as well. If you have a broody that wants to set but you don't need any more chickens, you should be able to place, say, some fertile duck eggs under her and she will set them as if they were her own. Though, keep in mind the size of the hen's body in relation to the size of the eggs. I have allowed a small bantam hen to sit on as many as four large fowl eggs and she was able to hatch them. Likewise, you may not want to stick small bantam eggs under a large fowl bird just because of the possibility of the large hen squishing the small bantam eggs.

When using a hen in a surrogacy role, also keep in mind the differences in fowl. It can be quite traumatic on a chicken hen when the ducklings she hatched for you decide to take a swim in the farm pond. A chicken hen has a hard time understanding why those cute little ducklings want so badly to get in the water. On the other hand, I know of folks who have used their broody hen to hatch turkey and goose eggs—both of which will quickly outgrow the hen. But in the case of turkeys, the hen can teach the turkey poults to forage for food, which they have a hard time figuring out on their own.

A good surrogate broody hen is your natural incubator, so don't be afraid to use her as such. But always give her time in between clutches to regain her strength and body weight before allowing her to raise another batch. A good hen can usually hatch and raise four clutches of chicks a year. If you incubate eggs in a homebuilt or commercially-produced incubator, you can also have your hen raise a portion of or all of the chicks that are hatched if it coincides with her own natural hatch.

We were fortunate enough to have a mixed-breed hen that came to us as an adult, so we did not know her actual age. She loved to go broody and was a terrific mother hen. She was a medium-sized hen, but was plump. She could easily sit on up to fourteen standard chicken eggs at a time. In one year, we allowed her to have five clutches of eggs and she hatched them all. As an experiment, after one of her hatches, I wanted to see how many chicks she would actually take as her own in addition to the ones she had hatched. Out of fourteen eggs she was sitting on, thirteen hatched and to that I added another fourteen chicks that hatched in an incubator the same day that she hatched hers naturally. So she had twenty-seven chicks that she cared for and she was just as content as could be. This is by no means a recommended practice but merely shows what a good broody can do. She was a fierce defender of her brood but allowed us to check her nest and handle her chicks as needed. Towards the end, she no longer laid her own eggs but she still went broody and eagerly hatched other hens' eggs for us. We actually had to break her from going broody so that she could rebuild her body weight before we allowed her to set again.

A good surrogate broody hen is your natural incubator, but always give her time between clutches to regain strength and body weight.

So, if you can find yourself a good broody hen, put her to work hatching and raising chicks or other fowl to either raise or to sell. Treat her well and she will reward you over and over again.

BREAKING A BROODY HEN

There are various reasons why you would want to break a hen of going broody. The most logical is just because you don't want any chicks, but there are other reasons as well that you would need to break a broody hen. Maybe because of her health or if she has not put back on sufficient weight since last going broody. Maybe you don't have a rooster and so the eggs are not even fertile. You don't need a rooster for a hen to go broody. Or maybe it is just not a good time of year. Hens generally go broody in the spring but they can go broody any time of year. Some hens are so broody that it seems that they are no more done with one batch of chicks and they want to go broody right again.

Whatever your reason, breaking a broody is not hard. It is a matter of cooling the hen's belly and keeping it cool. This is most easily accomplished by placing your broody hen in a wire cage elevated off the ground. Make sure she has feed and water and is protected from the elements, but do not give her any type of nesting material. Keep her in this cage for five days, and then release her back into her normal pen. If she heads back to a nest box and decides to just sit, place her back in the cage for a couple more days. If your weather is cool, she will break her broodiness much quicker than if it is hot out. The earlier you catch her sitting in a nest, the easier it will be to break her.

A man takes his brother to see a psychiatrist.

Psychiatrist: What seems to be the problem?

Man: My brother thinks he is a chicken!

Psychiatrist: How long has this been going on?

Man: About a year.

Psychiatrist: Why didn't you seek help sooner?

Man: Well, we needed the eggs!

MOVING A BROODY HEN

On occasion you might find yourself in need of moving a broody hen. This may sound easier said than done, but it can be done. You may need to move her because of fear of predation or to isolate her from the rest of the flock. Or maybe an egg has been broken in the nest or she is just nesting in a place that is not conducive to raising chicks, such as on a ledge or in a hayloft. There are a variety of reasons to need to move a hen.

This is not an easy task, as she will want to continue to return to her original nesting site. Once you have picked out a more suitable nesting sight for the hen, place a small cage around it that can be easily removed. In the dark of night, pick up the hen and the eggs and move them to the new nesting area. If you have to do it in multiple trips, move the eggs first and then the hen. Place the eggs in the nest, then gently place the hen over the eggs. Talk softly to her and pet her back until she relaxes and lies down on the nest. She may try to get up off the nest. Just place her back on the nest and pet her again until she lies down. She will eventually settle in.

Chickens cannot see well in the dark and, therefore, are less apt to want to run away. By having the temporary cage around her and the nest, if she does try to get up off the nest, she is limited to where she can go. It is fine if she is off the nest for a period of time. Once she calms down and sits back on the nest, she will adjust the growth of the embryos to compensate.

Once she seems to settle in and is back to her set, you should then be able to remove the cage from around her. This is usually the next morning. Make sure that she has food and water available to her. If she should decide to go back to her old nest at any time, you might find it necessary to keep the cage around her for the duration of her set.

You should never have to move a hen during the daylight except in emergency situations. If you find yourself in this predicament, be prepared for a fight from the hen—but the same guidelines apply. Find her a suitable spot, preferably dark and secluded, then prepare her a nest, move her and the eggs and place a cage around her. Her natural instinct is to get back to her eggs that she believes are still in the old nesting spot. She will just about drive herself crazy trying to get back to her old spot, but it is also her instinct to set. If the cage you place around her is small enough to limit her movements, she will eventually settle down and go onto the nest. This may take a few hours. If the weather is warm, there should not be any problem with her being off the nest for that length of time. If the weather is cold, you will need to find her a suitable nesting site that has some warmth to it to help keep her and the eggs comfortable.

Moving a broody hen is possible both in the dark and in the daylight. It just takes persistence and understanding from you to make it happen in the least traumatic way possible for both you and the hen.

All About Eggs

Chicken eggs are amazing things, from their creation in the hen's body, to laying and to storage and eating. They are incredible and worthy of a study all their own.

HOW AN EGG IS FORMED

Once a female chicken, a pullet, reaches sexual maturity, anywhere between seventeen and thirty-two weeks of age depending on the breed, her body will begin producing eggs, which she will lay on a varying schedule. How the egg is formed inside the hen is quite a remarkable thing.

Most females of the animal kingdom, including humans, have two functioning ovaries. In the world of chickens things are slightly different. Shortly after a female chick is hatched, the right ovary shrivels up and only the left ovary remains viable and produces eggs. This is not to say that the right ovary is lost forever. In rare instances, in the absence of a rooster in the flock, a hen's right ovary may regenerate to some extent, causing testosterone to be produced, and the hen will take on traits of a rooster and possibly even crow.

The reproduction system of a hen consists of five main parts, which together are called the oviduct: the infundibulum or funnel, the magnum, the isthmus, the uterus or shell gland, and the vagina. The ovary sits above the infundibulum and, in a newborn chick, contains the yolks of all the eggs the hen will lay in her lifetime. I know, you are looking at a chick and you look at an egg yolk and you are thinking I am crazy. The yolks within the ovary are nearly microscopic, being about the size of the point of a pin. These tiny starts of the yolk are called oocytes. These oocytes hang in clusters kind of like grapes do. It is estimated that in a standard laying hen there are approximately one thousand oocytes.

The whole process of laying an egg takes about twenty-six to twenty-eight hours. Some hens will lay sooner and many will lay later. For a good hen, you can expect to get five to six eggs in a row before the hen skips a day or two. This is due to the timing of the release of hormones that stimulate the laying process. These hormones are released during the night, usually sometime between midnight and 8 A.M. Ovulation takes place about six to eight hours later during daylight hours, and then the egg process takes twenty-six to twenty-eight hours, so in theory, a hen will lay her egg later and later each day during her cycle. Once the timing extends beyond one of these timing points, then the hen enters a period of rest until the next cycle starts.

When a hen is stimulated by nature to lay eggs, hormones are released in her reproductive system and a yolk will begin to grow. It takes approximately seven to nine days for the yolk to reach its full size. There are yolks of various sizes at any one time. Once a yolk reaches its full size, the hen's body releases additional hormones

Laying an egg takes about 26-28 hours. For a good hen, expect to get 5-6 eggs in a row before the hen skips a day or two.

that signal ovulation. When ovulation takes place, the yolk separates from the ovary and drops into the infundibulum, which is shaped kind of like a funnel.

In the infundibulum, the vitelline membrane is added around the yolk to help hold it together. It is also within the infundibulum that the yolk is fertilized if sperm from a rooster are present. The total time the yolk spends in the infundibulum is approximately fifteen minutes. From there, the yolk is funneled into the magnum, where the layer of egg white, which is called the thick albumen, is added. As the egg travels along it rotates, and through this twisting motion, fibers of the albumen form the chalazae, which are the two white stringy objects you see at each end of the yolk when an egg is cracked open. These two chalazae are twisted in opposite directions and hold the yolk centered in the egg. It takes the egg approximately three hours to pass through the magnum.

The egg continues on its path into the isthmus, where the two shell membranes are added. More albumen is also added at this point. This second layer of albumen is much thinner in consistency and can pass through these membrane layers and plumps up the egg. This process takes about one and a half hours.

After the membranes are added and the egg is plumped, the egg enters the uterus. It is within the uterus that the eggshell is actually formed. The actual shell is made primarily of calcite in the form of crystals of calcium carbonate. The egg remains in the uterus for approximately fifteen to sixteen hours while the calcite crystals grow and cover the egg in shell. The last layer of shell to be applied is the color layer. This is a very thin layer and the color is unique to each individual hen. It is kind of like her own fingerprint.

If you look at a shell under a magnifying glass, you are able to see that the egg is covered with small pores. These pores are a result of the growth of calcite crystal where they come together and fuse. These holes are very necessary to an egg's well being, whether it is for hatching or for eating.

The pores of the egg allow for the transmission of oxygen in and carbon dioxide out, which is especially important if you have a growing embryo. But the pores can also let in bacteria and contaminates. It is the shell membranes that block these foreign substances from getting to the yolk.

Once the shell has formed, the egg then enters the vagina. Throughout the egg-making process, the egg travels through the oviduct pointed end first. Once the egg enters the vagina, the egg turns around and will be laid large end first. As the egg passes through the vagina, it is also coated with a thin film that is made up primarily of protein. This thin film is called the bloom. This bloom is the first line of defense to keep bacteria from entering the egg through the pores. Once you wash the egg for eating purposes, the bloom is removed. Therefore, it is extremely important that eggs are refrigerated shortly after washing.

The fully formed egg passes through the cloaca, which is a tubular organ located just inside the vent where the digestive, reproductive and urinary canals all meet. The egg then exits the hen's body through her vent. When the egg is laid, the temperature of the egg is approximately 104ºF (40ºC). The egg quickly cools, and in doing so, the two inner membranes separate and an air cell is formed, usually on the large end of the egg. This is one of the ways to tell how fresh an egg is; the smaller the air cell, the fresher the egg. The longer an egg sits, the larger the air cell will grow due to evaporation of the contents of the egg.

The egg is a remarkable part of nature and deserves a book in and of itself. It is amazing to think that the reproductive system of the chicken can take a yolk and in just a few short hours can give you a beautiful egg and repeat this process over and over again.

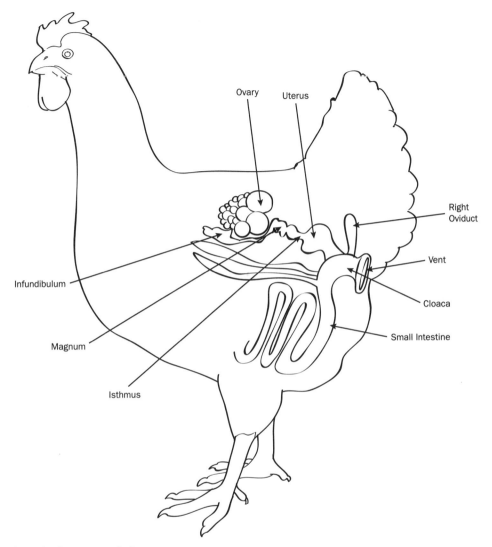

Reproductive system of a hen

> You know you're addicted to chickens when . . . you keep a baby monitor in your coop so you can hear if anything is going on at night.

COLOR VERSUS TASTE

Brown, white, green, blue. A rainbow of colors can be had in every hen house. Such a simple thing, yet so much discussion and controversy surrounds it. Which egg is healthier? Do they taste different? How long can they keep? How long does it take for a chicken to lay an egg? And on and on.

But where do you start? The common chicken egg comes in many different colors and sizes. And contrary to some people's belief, the feather color of the chicken does not have anything to do with the color of the egg that they will lay. Yes, there are white chickens that lay white eggs. There are brown chickens that lay brown eggs, but there are white chickens such as the White Rock that lay brown eggs and the Brown Leghorn that lay white eggs.

If you crack open an egg and pour out the contents onto a flat surface, you will see some very notable aspects of the egg. The most obvious is the yolk and the white, which is actually called the albumin. The white stringy things on each side of the yolk are called chalazae. These cords are elastic in nature and are what hold the yolk centered in the egg. Many people think that these are the start of a chick. They are not; they just hold things in place.

If you look closely at the yolk you will see a small light-colored spot about the diameter of a pencil lead. This is what is called the germinal disc. If the egg were fertile, this is what would develop into the chick—which brings up another point. How do you tell if an egg is fertile or not? If you look at the germinal disc and it looks like a bull's-eye, with an extra ring around the center spot, then it was a fertile egg.

On occasion you will see a dark spot in your egg when you crack it open. This is what they call a blood spot or meat spot. These are caused by a blood vessel breaking at the time the egg is first forming. They are completely harmless and can be simply removed before cooking. These blood spots are one of the main reasons the commercial egg producers use white eggs. It is easier to see a blood spot in a white egg than it is in a brown egg, and so the egg can be discarded before the eggs are sent to market.

Different breeds lay different colors of eggs and they can lay them in various shades of color. You really do not know what color of egg you will get until the hen starts into full production. The egg color can change as the hen gets older, and also due to diet. We raise Delaware breed chickens, as an example, and each hen's eggs vary in color from a red to brown to beige. We also raise Ameraucana breed chickens and their eggs can be blue or green. But the main question that arises is, "Do they taste different?"

There is absolutely no difference in taste between the colors of eggs. A chicken's diet can change the taste of an egg, as can the age of the egg. There is a huge difference in the taste of a store-bought egg versus a farm fresh egg. Have you ever hard-boiled a store-bought egg and a farm fresh egg and then tried to peel both of them? The store-bought egg will peel very easily while the farm fresh egg will peel apart in chunks. This is because a store-bought egg has sat in a warehouse someplace for up to three months before it ever made it to the store shelf. This allows the egg contents to partially evaporate and the shell lining pulls away from the shell, making it easier to peel.

Crack open a store-bought egg onto a flat surface and then crack open a farm fresh egg beside it. The farm fresh egg stands up taller, is much firmer and is usually a richer color than a store-bought egg. This is because the one is fresh and the other has been sitting around getting old. So naturally there is going to be a difference in taste.

Your chickens' diet also affects the flavor of the egg. If you feed your chickens a basic diet of cheap feed, the eggs are going to taste different from those fed a balanced diet of quality feed and supplements. If you feed your chickens leftovers from last night's dinner, this can also affect the taste of the eggs. It is recommended to not feed your chickens such things as onions or garlic as it can taint the flavor of the egg. Diet will also play a factor in the color of the yolk. Different feed additives and supplements, or the lack thereof, cause the yolks to be various colors, from a rich orange to a pale yellow.

The color of the eggshell does not, in any way, affect the taste of the egg. A green egg tastes just the same as a brown egg, as does a white egg or a blue egg. Eggshells are all white inside. In fact, you can take a brown egg and if you are careful enough and also have lots of time on your hands, you can scrub it enough to actually make it white.

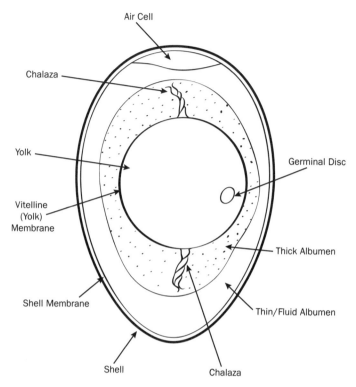

Parts of an egg

The big taste controversy started years ago with the advent of the big commercial laying houses. Before commercial egg production came into effect, it was simply just a matter of color preference, but once commercial production started, the growers found that white eggs sold better because they looked cleaner, and as millions of eggs were produced they also had to be stored prior to going to market. This is where the taste difference came in. I could cook you a fresh brown egg, a fresh white one, a fresh blue one and a fresh green one, and blindfold you and you would not be able to tell the difference in taste. But cook a fresh egg and a store-bought egg and yes, you would be able to easily tell the difference.

But which is healthier? All colors of eggs can be equally healthy. Through modern technology, commercial growers have been able to manipulate the healthy aspects of eggs. Some have lower cholesterol while others have higher levels of Omega-3s. It makes you wonder what they are actually feeding the chickens to be able to alter the contents. But in the normal realm of home production, one color of egg is just as healthy as the other. Again, it depends on the chicken's diet.

EGG FRESHNESS AND STORAGE

Many times I am asked about how long an egg can sit in a nest or how I know an egg is good to keep. This depends a lot on your conditions. The condition of the nest boxes, the outside air temperature and how you store your eggs.

It is recommended that you collect your eggs at least once a day, or more often if it is hot outside or freezing cold. It also depends on how many hens are trying to use one box.

Every egg has an air cell in it. Eggshells are porous and over time the contents of the egg will evaporate out and let air in, so the air cell grows. The older the egg, the larger the air cell.

After an egg is laid, it is then up to you to do with it as you wish. If you are going to incubate the egg, you will want to let it cool to ambient air temperature before adding it to the incubator. An egg that has just been laid is approximately 104°F (40°C). The egg must cool and then be brought back up to hatching temperature in order for the embryo to develop. Eggs for incubation should be clean and unwashed. These eggs can be stored in a cool, dark place for up to ten days before fertile viability starts to fade.

If you are going to eat the eggs, it is best that they are gently washed, completely dried and placed in the refrigerator. But this brings up an interesting point. Prior to the advent of refrigeration, eggs were just stored on the counter in a wire basket. How was this possible? As mentioned above, when an egg is laid it is covered with a very thin film called bloom. This bloom protects the egg from bacteria getting inside. An eggshell is actually very porous. If you look very closely at the shell, you will see many little holes in it. This allows moisture to escape from inside the egg and also lets air in. The bloom helps keep the bad bacteria out. It is said that a clean, unwashed egg can remain good for up to seven months without refrigeration if the ambient air temperature is in the 65°F (18°C) range. Personally I would not try eating an egg that sat on the counter for seven months, but it is not unusual for eggs to sit on our counter for up to two weeks during the winter months. Once an egg is washed, it needs to be refrigerated right away.

The longer your eggs sit out, the more the contents will evaporate. Many people will allow their fresh eggs to sit out so that they become easier to peel if hard-boiled. If you hold a strong flashlight up to an egg so that you can see through it, you will see a very small air cell at the fat end of the egg. The older an egg gets, the larger this air cell expands. In an egg set aside to hatch, this air cell will grow as the chick develops. A day or so prior to the chick hatching, it will use its beak to poke a hole in the membrane separating the air cell and take its first breath. In an eating egg, this air sack will grow in size depending on the age of the egg. An old time anecdote for testing the freshness of eggs was to place eggs in a bucket of water. If the eggs sink, they are fresh. If the eggs float, they are old and should be discarded.

Dirty eggs should be discarded right away. Eggs should also not be allowed to sit in a nest for more than twenty-four hours, shorter if there are multiple hens using the same nest. What happens is that other hens coming into the nest can and will scrape the bloom on the eggs that are already in the nest. When this happens, it allows bacteria to more readily enter the egg. A hen may have poop on her feet, which will get on the eggs. Sometimes you will have a hen that actually poops in the nest right after laying her egg. This is usually an older hen, and she should be culled or otherwise removed from the laying flock. Your nests should be clean and nesting materials changed on a regular basis. Nesting materials also need to be dry at all times. Wet nesting materials will allow the bloom to be washed off the egg and, therefore, let in bacteria.

People worry about such diseases as Salmonella and E. coli when it comes to eggs, and rightfully so. These are both results of dirty eggs or improperly handled eggs. Salmonella is much more common in eggs than E. coli. Both can be prevented with careful handling of your eggs as well as making sure that you completely cook your eggs before eating.

Now, just because store-bought eggs sit around for months before you get them doesn't mean they are not fresh; they are just not as fresh as a farm fresh egg is. Your store-bought eggs are kept in strict climate controlled facilities whereas your farm fresh eggs are stored by whatever means you choose.

Think of your grandmother or even your great-grandmother at a time before refrigeration. How did they manage to keep their eggs fresh? First off, eggs were a main staple within the household so many did not sit around for long. But the farmer's wife also knew that when wintertime came, there would be fewer eggs available.

Clean, unwashed eggs were just stored on the counter in an egg basket. Dirty eggs were usually fed to the hogs. For those farmers who had many hens and a great supply of eggs, they would talk with the town's pharmacist and order a product called water glass. The eggs would be put into barrels or crocks and then the water glass would be added to cover the eggs. This allowed them to keep the eggs for months and, therefore, provide eggs right through the winter.

Without modern technology and fancy testing equipment, there is very little you can do to make sure an egg is fresh. About the only way that you can test the freshness of an egg is with a water dunk test.

To test an egg for freshness, get a tall glass and fill it with water. Lower the egg in question into the glass. If it sinks to the bottom and lies on its side, it is the freshest. If it floats in the middle of the glass, it is somewhat fresh. If it floats at the top of the glass, it is definitely not fresh and should be tossed out. This can also be done with a large amount of eggs in a large bucket. As touched on before, the older the egg, the larger the air cell and, therefore, the more they will float.

But, like with anything, if in doubt, throw it out. Also be aware if you're going to water test your eggs, they'll need to be refrigerated after because the water will wash the bloom off the egg, making it more susceptible to bacterial contamination.

You know you're addicted to chickens when . . . you call home from work to ask how many eggs have been laid, what color they are, and which nest box they were in.

 ## Incubating Eggs

So, you think you might want to incubate your own eggs or maybe you are in the process of incubating a clutch of eggs. Well, here is a subject that you will get a hundred different answers to each question that you might have. We will handle this in as diplomatic a way as possible.

Each person has their own way of incubating eggs and each situation is totally different from the next. There are so many variables involved in the incubation process that it would be impossible to comment on every situation. We will try to cover as many different aspects as we can.

The best incubator that you can get is simply a good broody hen. Nature knows best when it comes to these kinds of things, but not all hens go broody, and depending on how many chicks you would like to have, not all hens can handle enough eggs. So we turn to incubators to enhance the process.

People will tell you that you need the biggest and the best incubator to be able to hatch eggs. They will tell you that one model is no good while another is the best. The truth is, there are many different styles and sizes of incubators in all price ranges. What matters most is what works for you and your situation.

People have been incubating eggs for hundreds of years. What did they do before all these new fancy digitally-controlled, automatic load-it-and-leave-it incubators became available? They used their imagination and the knowledge of hatching to compensate for the lack of available broody hens. All you need for incubation is fertile eggs, a reliable heat source and a thermometer. Granted, the more you are able to regulate the environment that the eggs are kept in, the better your hatch rate will be. My mother remembers her father placing eggs on the door of their wood stove and hatching chicks. When I got my wife involved in the whole chicken thing, I showed her that hatching was easy. I took a large Pyrex bowl and placed a basket in it. I put a bit of water in the bowl and placed a heat lamp over it. I placed four eggs in the basket and twenty-one days later we had two healthy, happy chicks hatch out. A 50 percent hatch rate from a bowl is not too bad. So it goes to show you that you do not need the newest and fanciest incubator on the market to be able to hatch a few chicks.

SOME BASIC RULES

There are a few basic rules to incubating eggs. First, throw out 90 percent of what you have read. Like I said before, everyone has their own opinion and you will get a hundred different answers. Your hatching situation will be different from your neighbor's. And his will be different from ours. You have to establish what works best for you and your environment. There is a lot of trial and error. The method you use when hatching eggs will depend a lot on what type of eggs you will be hatching. Chicken eggs are different from turkey eggs. Quail eggs are different from pheasant eggs. For simplification, we will be discussing the hatching of chicken eggs throughout this guide.

The first thing you will need is fertile eggs. These can be from your own flock or you can acquire them from a different source. A hen will lay an egg without a rooster being present, but you must have a rooster if you want fertile eggs. And just because you have a rooster doesn't necessarily mean that you have fertile eggs. The rooster himself has to be fertile. Age and time of year play a big part in this. A rooster loses fertility with age and also during the winter months. A one- to two-year-old rooster during the spring will be much more fertile than a five- to six-year-old rooster during the fall and winter.

The second thing you need is to know that an embryo begins to form when the temperature of the egg reaches 88ºF (31ºC). The optimum temperature for a chicken egg to hatch is 99.5ºF (37.5ºC). Lower temperatures will delay hatching and will potentially cause deformities in the chicks. Higher temperatures will speed up hatching but will also potentially cause deformities. Temperatures above 104ºF (40ºC) will generally kill the embryo. So, for proper growth and viability, you will want to maintain a temperature as close to the 99.5ºF (37.5ºC) mark as possible.

Thirdly, you will need to roll the eggs. This keeps the embryo from sticking inside the egg as well as exercises it. A broody hen will roll her eggs up to fifty times a day. For incubation purposes, you will want to roll the eggs at least three times a day and more if possible.

Beyond these three things, it all becomes a matter of what works best for you to increase your hatch rate and viability of the chicks.

EGG VIABILITY

The fertile viability of your eggs will be one of the most important things for a successful hatch. If your eggs are not fertile, they simply will not hatch. But how do you tell if the eggs are fertile? First you have to have a fertile rooster. As stated earlier, there are other factors as well, such as the age of your birds and time of year.

The best way to test the viability of your eggs is to run a test set of eggs through your incubator and at day 10, candle each egg to see if there is growth (more on candling later in this chapter). You can also crack open a fresh egg to see if it is fertile. Do this by gently cracking open an egg onto a clean plate. On the yolk of the egg you will see a small light-colored spot. This is the blastodisc or germinal disc. This is also sometimes referred to as the blastoderm. Look very closely at this spot to see if you can spot concentric circles around this spot. This is what is called the "bull's-eye."

If you see an obvious bull's-eye, then your egg is fertile or at least has the potential of being fertile. If there is no bull's-eye, then your egg is not fertile. If you have multiple hens being serviced by one rooster, there is a chance that some of the eggs will be fertile while others will not be. Therefore, it is best to run a test batch of eggs through the incubator to see what percentage of your eggs is showing fertility.

Eggs also need to cool and rest before being placed in the incubator. Do not sit and wait for your hen to lay her egg and then rush to put it into the incubator. A newly laid egg needs to cool for the shell to completely harden and the bloom to dry. It also needs to rest for the yolk to center itself. Hatching eggs can be stored for up to ten days before being placed in the incubator. After ten days, the eggs begin to lose their viability. This is not to say that you can't take two-week-old eggs and place them in the incubator, it is just that the chance of the eggs successfully hatching goes way down. I have heard of people hatching eggs that are over three weeks old.

HEAT SOURCE

A reliable heat source and thermostat are very important for the incubation process. Most modern incubators use a heating element somewhat like the element in an oven. Antique incubators used such things as oil lamps for their heat source. What the heat source is doesn't matter so much as long as it is reliable and adjustable. In nature, a hen's body temperature is between 104°F (40°C) and 107°F (42°C). By cutting down on her feed consumption, pulling feathers from her belly and moving eggs, you can help her to regulate the temperature around the eggs.

As stated earlier, a chicken embryo begins to develop when it reaches a temperature of 88°F (31°C). Optimum temperature for proper growth is 99.5° (37.5°C).

The second part of this equation is temperature regulation, which is usually handled by some type of thermostat. Modern incubators have either a digitally-controlled thermostat or what is called a wafer thermostat. The closer you can keep the temperature to that 99.5°F (37.5°C) magical number the better. Your thermostat should be able to regulate the temperature inside the incubator so that there is no more than a +/- variance of 2 degrees.

When a hen is sitting on her clutch of eggs, she will constantly move the eggs around to regulate their temperature. If they are too hot, she will move them to the outside. If they are too cold, she will move them farther under her. On really hot days, you will many times see the eggs lying all around the outside of her body because she knows that it is too warm to have them under her.

EGG TURNING

Turning your eggs is very important. It exercises the embryo as well as keeps it from sticking inside the shell. You can turn your eggs by hand a minimum of three times a day. If you are incubating a large number of eggs, this can become very tiresome. Also, it means that you have to open your incubator each time, which causes you to lose heat and humidity and, therefore, causes the eggs to take longer to hatch. Egg turners do the work for you. They gently rock the eggs back and forth. A hen will turn her eggs up to fifty times per day. A turner will rock the eggs about six to eight times a day. If you are going to use a turner, place the eggs into the racks pointed side down. If you are going to turn the eggs by hand, before placing the eggs in the incubator mark an X on one side; this will give you a reference when turning. A turner will greatly improve your hatching experience.

HUMIDITY

You will hear and read a lot about humidity. Humidity plays a big part in the growth and hatching of chicks. Too little humidity and the chicks will stick inside the shell. Too much humidity and the chicks will drown when they pip. The general rule of thumb is that the humidity be kept around 50 percent for the first eighteen days of incubation and then 70 percent for the last three days.

There are many schools of thought on this and again it is what works best for you. For us, we hatch small batches of chicks in tabletop incubators. We find that dry incubation (not adding water to the incubator) works best for us, but we also live in Florida where there is lots of natural humidity. But we are not alone in this thought. Many people are starting to go to dry incubation and finding that they have better hatch rates. If you find, upon hatching, that the chicks are sticking to the shells, then you will need to have higher humidity. If the chicks are fully developing but dying just before hatching, then they are most likely drowning and your humidity is too high.

If you have ever picked a brooding hen off of her nest, you will feel that her belly is very warm and feels almost wet. Not only can she control the temperature under her but she can also, to some extent, control the humidity.

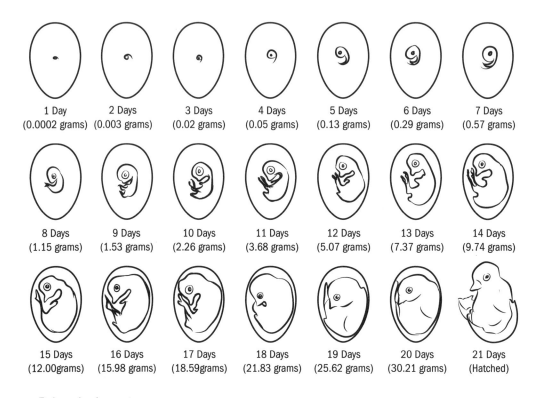

Embryo development

If you reach the point of hatching and find that the chicks are sticking inside the shells and having difficulty hatching out, you can lightly mist the eggs with warm water. The water will be absorbed through the pores in the shell and help moisten the inner membranes.

Again, I can't stress enough, that it is whatever works best for you and your situation.

CANDLING

Candling is the act of shining light through the egg to check the growth of the embryo. You can go buy one of the fancy egg candlers, but you can also do the job with a good flashlight.

Candling is usually done between day 7 and day 10 of the incubation process. It needs to be dark where your incubator is, so either candle at night or, if your room can be made dark other ways, then you can candle any time. The brighter and more

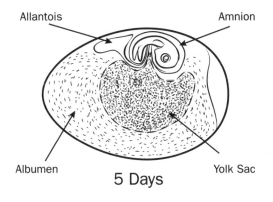

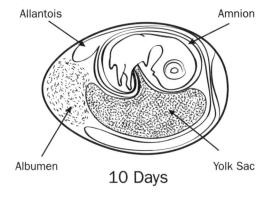

Daily changes in the weight and form of the developing chick embryo (White Leghorn)

concentrated the light beam is, the better for candling. We use an LED flashlight that has tripod legs. Hold the egg close over the light source and block any light coming around the egg. At around day 7 to day 10, you should be able to see lots of veins coming off of a central spot. This will indicate that the embryo is growing and viable. Place the egg back into the incubator and move on to the next. If the egg appears clear, then the embryo did not start developing for any number of reasons and these can be tossed out. If there is a distinct dark ring around the inside of the egg, this is what is called a blood ring and it indicates that the embryo has died and these eggs should be discarded as well.

Many people will also candle their eggs at day 18 to make sure of proper growth. We do this as well. You should be able to see a dark mass that takes up most of the egg and also a clear area at the large end of the egg that is the air cell. If your light is bright enough, you may also be able to see the chick moving around inside the egg.

LOCKDOWN

You will hear the term lockdown a lot. Lockdown is the period of time from day 18 to day 21 when you stop turning the eggs, raise the humidity and wait for chicks to start hatching. This is the hardest time for most people because they are eagerly awaiting the hatching of the chicks. The moments of the last eighteen days go through their heads wondering if they did everything correctly. They want to know how many of the little fluffy chicks will soon be running around. Take a deep breath and wait patiently. Keep your temperature steady and give the eggs a bit of extra humidity.

We incubate our eggs in small incubators, then transfer them to a home-built hatcher for lockdown. We place a small glass of water inside to help raise the humidity just a bit and find this works well for us. We also allow our eggs to stay in the incubator for nineteen days instead of the customary eighteen. That is unless we are hatching lots of eggs and need the incubator for another hatch. We have found that this extra day in the incubator has not adversely affected our hatch rates.

HATCHING

The day has arrived. You have kept the temperature at 99.5º (37.5ºC). You have rolled the eggs at least three times a day. You candled the eggs and they all looked good. You went into lockdown and waited patiently. It is now day 21—where are the chicks? The chick will first poke through the inner membrane of the egg into the air cell and take its first breath. It will also get into position to start breaking free of the shell. If all went well, then you will soon see what is called a pip. This is a small hole that the chick breaks in the shell to be able to get outside air. The chick has what is called an egg tooth on the tip of its beak that allows it to more easily break through the shell.

Your chick now may rest for up to twenty-four hours before it starts to make it's move, but generally after just a couple of hours, the chick will begin to "zip" the shell. This is the process where the chick breaks a seam around the shell so that it can get out. A chick will always zip the egg in a counter-clockwise fashion if you are looking at the large end of the egg. This is very hard work for a little chick and it will rest many times along the path. This process will all depend on the chick as well as on the slickness of the inside of the egg. Some will take hours and others will be done in just a few minutes.

Once the chick has zipped a line around the egg, it will push the shell halves apart until it is free. The little chick is exhausted at this point and will want to sleep as much as possible.

AFTER THE HATCH

You are giddy with excitement to see all these wonderful little fuzzy bodies laying there. So, now what? The chicks will be wet from fluid from inside the egg. Leave them be for about twenty-four hours until they get all dry and fluffy. They will be up walking around and then lie down to sleep. They don't have a whole lot of energy yet and what they do have is used up really quickly. Once the chicks have dried, they can then be moved to a brooder. Your brooder should be draft free but have good ventilation. It should be at 95ºF (35ºC). Place some paper towels on the bottom so that the chicks don't slide around when trying to walk. You can also sprinkle some chick starter feed on the paper towels so that they can learn to eat. Do not be alarmed if they do not eat right away. A chick can survive for up to three days just on the yolk sac that was absorbed into their body prior to hatching. Have water available to the chicks. Place marbles or small stones in the dish of the waterer to help prevent the chicks from falling in and drowning in too-deep water; these can be removed when the chicks are about 5 days old. Some chicks will figure out the water on their own. If on day 2 you do not see them drinking on their own, take each chick and gently dip its beak into the water and then lift its head back up. This will let them know where the water is.

Soon, your little chicks will be running around eating and drinking and bringing you hours of enjoyment. After a couple of days you can remove the paper towels and allow them to be on clean litter. Be sure to keep their waterer full, their bedding clean and the temperature right, and you should have some happy little chicks. The brooder temperature can be dropped five degrees Fahrenheit each week until you reach ambient outdoor temperature. Be on the lookout for any signs of distress. If your chicks are huddled together, then they are too cold. If they are all up against the edges of the brooder, then they are too hot. If they spread all out, then they are just right. Illnesses will be covered in a different section.

Date	Day	Action	Temp	Humidity	Turn Eggs	Comments
	0	Pre-Heat Incubator			—	
	0	Set Eggs				
	1					
	2					
	3					
	4					
	5					
	6					
	7					
	8					
	9					
	10	Candle Eggs				
	11					
	12					
	13					
	14					
	15					
	16					
	17					
	18	Eggs Lockdown			—	
	19				—	
	20				—	
	21	Hatch			—	
	22					
	23					
	24					
	25					
	26					
	27					
	28	Discard Eggs				

Calendar to keep track of the incubation process

KEEPING RECORDS

I advise people to keep a daily log of their incubation process, especially if they are new to incubating their own eggs. This is most easily accomplished on a simple calendar. Mark the date when the eggs went in. If you set your eggs in the morning, that day counts as day 1. If you set your eggs in the afternoon or evening, then the next day counts as day 1.

Mark your candle date, your lockdown date and your tentative hatch date.

Each day, keep notes of temperature, humidity and anything that happens, such as temperature spikes or drops. This will give you a reference to look back on if you should have eggs that go to your hatch date but don't hatch. At the end of your incubation, look at the eggs that didn't hatch. Crack them open and see at what point in the growth cycle they died. This will allow you to better control your incubation in the future.

DOS AND DON'TS

There are many things that can potentially go wrong with a hatch and I will try to touch on some of the more common ones.

Before placing your eggs in an incubator, allow the incubator to come up to temperature for at least twenty-four hours. This will allow you to get the temperature more closely regulated. If you are using an automatic turner, also turn this on as the turner motor produces heat and will affect your temperature settings.

Place your incubator in a draft-free area that has a pretty regulated temperature. This will make it much easier for you to regulate the temperature inside the incubator. Do not place your incubator under a ceiling fan, near a drafty window or door or near a heating and cooling vent. Nor do you want to place your incubator where it can be affected by sun through a window.

Hatching eggs can be stored for up to ten days prior to incubation. Store them in a cool place that is about 65ºF (18ºC). Beyond ten days, the eggs lose much of their viability. This storage period allows you to collect enough eggs for your hatch.

Use only clean eggs, but do not clean the eggs. Each egg, when the hen lays it, is covered with what is called bloom. This protects the egg from becoming contaminated by outside sources. If you wash the egg, this can allow bacteria to enter the egg and cause it to become rotten. Make sure your hens are laying in clean nest areas.

Allow your eggs to rest for at least twenty-four hours. You do not want to stand there waiting for your hen to lay an egg and then rush it to the incubator. An egg actually needs to cool below 88ºF (31ºC) first and then be brought back up to temperature for proper hatching. This time also allows for the yolk to re-center itself in the egg if the eggs have been shipped or moved any distance.

Place your eggs correctly. If you are using an automatic egg turner, place your eggs pointed end down in the racks. If your eggs are oval, shine a strong light through the egg to see which end the air cell is on. The air cell goes up. If you are going to turn the eggs by hand, mark an X on the side. This will give you a reference point when rolling the eggs. Roll half way one time and roll back the next. Also mark on the end of the egg the breed of chick if you have more than one breed that you are trying to hatch. Many chicks look very similar at hatch.

You can also mark your eggs with the date it was laid. Try not to use a permanent marker for marking your eggs. Many permanent inks will leak chemicals into the shell that are potentially hazardous to the growing embryo. A crayon or pencil work fine.

You should only open your incubator when absolutely necessary. If things are going well with your incubation and you are using a turner, then you should only need to open your incubator to place eggs in, add or remove water for humidity adjustment, to candle, at lockdown to remove the turner and after the hatch. You will hear many people groan about opening an incubator, but if you have been around poultry enough you realize that the hen leaves the nest at least once a day to eat and drink. This would be no different than opening the incubator once a day, but really, there should be no need to open it unless you are turning your eggs by hand. This constant cooling and heating of the eggs would delay your hatch a bit.

UH OH, WHAT HAPPENED?

Sometimes something unexpected happens. Here are a few scenarios I've come across, and what you should do.

An egg explodes.

This does happen from time to time and it makes a stinky mess. An egg will explode if bacteria get into the egg and cause it to rot. This generally happens if you are using dirty eggs. If this should happen, you need to immediately clean your incubator to help avoid further contamination of the other eggs. Gently wipe off the surrounding eggs and place them in an egg carton or in a towel to protect them while you are cleaning up the mess. Work quickly to clean the incubator and then place the remaining eggs back into the incubator to continue the process.

I bought eggs on eBay and just candled them and none of them seem to be growing.
This is a common problem. I am not picking on eBay in any way; I am telling you that shipped eggs are not a reliable source. The yolk of an egg is held in place on each end by what is called chalazae. These are delicate cords that keep the yolk centered in the egg. When you crack open an egg, you notice a white stringy thing on the yolk: this is the chalazae. When eggs are shipped, they encounter many things that are not favorable to their well-being, such as postal handlers that toss the packages, machinery that sorts the mail, bumpy vehicle rides and such. By the time the eggs get to you, they are pretty much scrambled inside. So, if you are going to buy eggs and have them shipped to you, be aware that the viability could very possibly drop to near zero. Yes, there are those rare instances when they arrive safely, but it is always a gamble.

It is day 21 and the chicks are not hatching.
Take a deep breath. Just because it is day 21 on your calendar does not mean that it is day 21 of the growth process. Many factors come into play here. Was your temperature exactly accurate? Probably not. Every time your temperature drops below 99.5ºF (37.5ºC) it slows the growth of the embryo. When you placed the eggs into the incubator, they were cold and had to be brought up to temperature. It takes some time for the mass of the egg to reach that 99.5ºF (37.5ºC) magical number. How many times did you open your incubator? This cools the eggs each time it is opened. We have had eggs hatch as late as twenty-seven days after placing them in the incubator. We use the rule of thumb that if the eggs do not hatch by day 28, then they are probably not going to hatch. This gives an extra week just in case.

The chick has pipped but is not doing anything.
This is normal. This is a huge struggle for the chick and it gets tired very easily. Think of it as if you were sewn into a sleeping bag and had to chew your way out without using your hands. If a chick has pipped and has done nothing for more than twenty-four hours, we make sure the chick is still alive. We take the egg and

hold it tight to our ear and gently tap on it with our fingernail. If the chick is still alive it will usually chirp or you can hear scratching.

The chick started to zip the shell but has quit.
Again, this is a tough process and the chick will stop to rest, but it also brings up the point of "should I help it?" Generally, no! This is a really tough call that can end up with tragic results. The chick has to absorb all the yolk sac before it can come out of the shell. The chick knows when that point is. It is not unusual for a chick to pip the egg and even partially zip the egg and then rest while the remaining yolk is absorbed into its system. If you help the chick and the sac has not been absorbed, or there is still external veining, then you risk killing the chick. But if the chick is stuck and cannot turn to complete the zipping because the membranes of the egg have dried to it's fluff, then the chick is at risk of dying from exhaustion. So it becomes a tough call of when to help and when not to. We have found that if we watch the chick and the membrane around the zip line is really drying out and the chick does not seem to be able to move, then we will take a toothpick and ever so carefully pick the shell a bit along the zip line. We will do this in small stages. Watch for any veining or blood. If this is encountered, stop immediately. Only work on the large end of the egg, because if there is still yolk left it will be at the small end of the egg. Many times, just a small chip will free the chick to be able to move again and it will continue on its own.

A chick absorbs the yolk sac, then leaves the shell.

My chick hatched, but it was all slimy and now it is drying into a hard clump.
This does happen from time to time. If too much of the membrane came out on the chick, it will rapidly dry and not allow the chick to fluff. Dampen a soft washcloth with warm water and gently wipe the chick from head to tail to help remove this gunk. It may take a few times to get it fairly clean. Remember that chicks chill quickly, so use warm water and work in stages.

My chicks did not all hatch at the same time and some are running around while others are still just pipped.

We do not hatch or brood in our incubators. We have a home-built hatchery that the eggs are transferred to at lockdown and we also have the brooder all set up and warmed by the time the eggs start to hatch. Chicks can survive for up to three days without food or water, but if you are prepared for them they should be just fine. We do things a little bit unorthodox when it comes to hatching but it is what works for us. At day 18 or even day 19, the eggs are transferred to the hatcher. This opens up space in our incubators for the next round of eggs. It also keeps our incubators cleaner. As the chicks hatch, we remove the empty shells and allow the chicks to dry and fluff, which usually takes about four to six hours. If they are walking well, we then transfer them to the brooder. This keeps them from pecking at the ones that are still hatching and also gives extra room in the hatcher. Many people say, "No, no, you can't do that. They are on lockdown and you can't open it up." Well, this is what works best for us and we do not lose chicks because of it.

Hatching chicks is a fun and rewarding adventure. Just remember that what works for one person does not necessarily work for another. There is a lot of trial and error involved and good note keeping does help. Chicks are very resilient to many things. Many times it is just a matter of taking a deep breath, assessing the situation and looking for the most logical solution. Hatching chicks is very easy and should not be made troublesome just because of a few people's misadventures. There will be death involved, as very few people ever achieve a 100 percent hatch rate, but if you can achieve a hatch rate of 75–80 percent, then you are doing well. Even the huge hatcheries do not achieve a 100 percent hatch rate. If you have a complete failure in the process, look back on your notes to determine why and make the necessary changes. Most of all, enjoy yourself and the adventure you are on.

FALL CHICKS

I am a huge advocate of raising fall chicks. With a bit of careful planning, you can be also. The hardest part is the availability of the chicks.

I know, you are saying, "Are you crazy? How can we rear up a flock of chicks when there is snow on the ground and the winds are blowing at 40 miles per hour?" Well, anything is possible with a little ingenuity. Let's first look at the logic behind it.

Mother Nature tells us that chicks hatch in the spring. After all, this is when the hens are the broodiest and that is just the way that nature intended. The hatcheries all start selling chicks around February and March and quit around August. The hens and roosters are at their most fertile in the spring. But we can go against Mother Nature.

Let's say you buy chicks from a hatchery in February and they show up on your doorstep the first day of March as day-olds. You spend the next five and a half months feeding and caring for these chicks to get them to the point of lay, which for most breeds is right around the twenty-two-week mark. Some will be sooner, but then some will be later. So, we will use twenty-two weeks as an average. By the time your chicks are old enough to lay, it is now the middle of August. Your young pullets will lay a few eggs here and there as their systems develop and get into production mode. The eggs are still on the small side and it is hot outside. It takes about three months for a young pullet to get into full production mode. So as their systems are developing we are now getting into fall and then into winter. The days are getting shorter and the weather is getting colder. A hen requires a minimum of fourteen hours of daylight to maintain production. Your girls will also go into their first molt, which will slow or stop their egg production.

Now we are into the throws of winter. The snow is piling up and the winds are howling. It seems that your day no more than starts and it is already dark again. The girls are not thinking about laying. You feed them now all the way through to February with just a minimal amount of eggs being produced. The snows are starting to melt and the temperatures are starting to rise again. Your girls are now just over a year old before they start into their first full year of production.

OK, now look at it the other way. Say you get your chicks on August 1. It takes them approximately six weeks to feather out, which puts you into mid-September. The weather is just starting to cool. The weather turns cold around the end of October on average, which puts your birds at about twelve weeks. You have a nice, secure, covered area for them out of the wind and elements. You raise them through the winter—what else are you going to do when it is nasty outside? February rolls around and your birds are now six months old. For the last couple of weeks they have been laying an egg here and there. The springtime temperatures are coming back and soon your girls are in full swing. You see your first full year of production six months before everyone else. That is six months less that you would have to feed them without much return.

But how do you work around the logistics of acquiring chicks in the fall? There are some hatcheries that will have a limited number of chicks available that late in the season. These would be your standard breeds such as Rhode Island Reds, Buff Orpingtons and Leghorns. If you have your own established flock, you could hatch eggs from them in order to perpetuate your flock. This would need to be done with an incubator, or if you have a very broody hen, then she could hatch for you.

Here in the South, we actually start our hatching season in October and run non-stop right through until June. And yes, it does get cold here. It is not unusual to have temperatures down into the teens. Our biggest problem lies in fertile egg availability. Our breeding birds are run under artificial lighting during the winter months so that egg production is kept up. We hatch eggs in small incubators inside the house and the chicks are brooded for the first two weeks in our office. After the two weeks, the chicks are transitioned to an outdoor pen that is equipped with a heat lamp and they are kept under heat until they are fully feathered at about five weeks.

But again, we do things a bit unorthodox around here. We work hard to acclimate our chicks as quickly as possible. When brooding chicks, the standard is to have the temperature of the brooder at 95ºF (35ºC) and then drop the temperature five degrees Fahrenheit for every week of age so that by week six the temperature is at about 70ºF (21ºC). That is fine if it is the middle of summer but you would not want to take six-week-old chicks that are at 70ºF (21ºC) and expose them to 40ºF (4ºC) temperatures right away. Most would probably die.

Instead, we let the chicks tell us the temperature at which they are comfortable. We raise the light as much as possible each day just so the chicks do not huddle together and try to squish each other. We do this so that by week three the chicks are just under an 85-watt floodlight and by week six they are comfortable with the ambient air temperature. We find that by doing this, the chicks feather out faster and are much healthier.

The real key to raising fall chicks is to have them in a place that is free from drafts. A corner of the barn or a shed works great for this. You can close it off as much as possible so that the temperature can be easily regulated. Put a good layer of litter on the floor to insulate them from the cold coming up from underneath and a heat source above to help keep them warm. Once the chicks get fully feathered, you could transition them to the chicken coop. If you live in the North, your coop should be well built to withstand the snow and winds and sealed from the elements.

The chicks should be fine in the coop. You may have to provide a supplemental heat source just on the coldest of days. On the nicer days, the chicks could be allowed to roam in an outside pen.

By raising fall chicks, when spring comes around and your neighbor tells you that he is going to order his chicks, you can just smile at him and tell him that you are waiting on your first egg because you have just spent the winter raising yours.

4 MAINTAINING YOUR FLOCK

Hierarchy of Poultry Society

The Poultry Society is a dynamic hierarchical society—there is one lead rooster and one lead hen, followed in succession by the others. We generally refer to this as "the pecking order." From just a few days old, chicks will begin establishing a pecking order within their group. This is who eats and drinks first and how they follow in line down to the weakest bird. As the chicks get older, they will begin sparring with each other to establish dominance. As they reach what we call the teenager stage, they can actually begin fighting amongst themselves, and as they reach the adult stage, if left on their own, it would not be uncommon for them to fight to the death.

Early on, during the chick stage, a young cockerel will display his dominance over the other chicks. As he and the others grow, this dominance will be challenged. If he wins this challenge, he remains the lead rooster; but if he fails, then another will take his place. The same holds true for the young pullets, but it is a much subtler approach. A lead rooster will control the flock. It is his duty to provide protection from predators by clucking and growling to warn the others of pending danger. It is also his duty to find food and water and to mate with the hens to perpetuate the flock. At any time, this lead role can be challenged and this includes by you. Chickens live in societal groups. Each chicken can recognize up to one hundred members of a flock. If you have more than one rooster, then each rooster will have control over his part of the flock. They will have their own territories, but these societal groups can overlap, and within a large flock the members of each sub-flock can change.

If you have an established flock and add another flock to it the new flock will stay to themselves and try to establish their own territory. If the new flock has a rooster in it, the established rooster and the new rooster will fight for control of territory.

It is very possible to have more than one rooster in a flock. Each rooster can easily take care of ten to twelve hens. If your poultry area is large enough and there are sufficient hens for each rooster, then they will form smaller subgroups and stay more to themselves, but you will still have one lead rooster over all the sub groups.

Have I confused you yet? Let me see if I can make this easier to understand by using our flock as an example. Right now we have approximately 150 chickens of various ages. We have six breeding pens, each with their own rooster. We also have a community pen that has two older roosters as well as one young cockerel. We let out one breeding pen of birds each day to mingle with the other community pen birds until we get to breeding season. This means that there is a different breeding rooster mingling with the other two roosters and the cockerel each day. Now, by doing this, you would think that we would constantly have fights among the roosters, when in actuality I cannot remember the last time we had a true fight.

Each of the breeding roosters has his own hens to watch after when he is out. Of the community pen, the cockerel has a few young pullets that he hangs out with that the older roosters are not yet interested in because they are not laying. One of the older community roosters has not been able to establish himself within the flock as most of the girls really don't like him, so he just kind of roams around trying to impress the girls. Our other community rooster is a small Serama breed and he has his own little flock of hens that pretty much stay right by him.

So, we have all these little sub flocks within the main flock. There are also plenty of hens to go around. The breeding roosters each have up to eight hens. The one community rooster has about six hens. The young cockerel has five pullets. The wayward community rooster has none that pay attention to him and then there are about forty community hens along with a bunch of younger pullets.

In our combined flock, the lead rooster is the little Serama rooster. He controls the whole yard. He is dominant even over the larger, heavy-breed roosters as well as the breeding roosters. If another rooster gets out of line, he will run over and jump at him. Everyone pays attention to what he does. The second rooster in command tends to be whichever breeding rooster is out that day. Then comes the other community rooster followed by the young cockerel. Of the hens, we have two older girls that are in charge, and the rest of the girls follow. This dynamic is challenged on a regular basis but rarely ever changes. The only changes to the order are in the lesser roosters and hens.

I hope that this made it easier for you to understand how a pecking order works.

But how is it controlled within the flock? Chickens have twenty-four known vocalizations and each breed has its own range of vocalization. So the sounds your flock makes can be extremely varied. A rooster crows for various reasons: mainly to communicate with his girls and also to establish his territory. The notion that a rooster only crows in the morning because the sun is coming up is purely a myth. Roosters crow throughout the day and also during the night. The more roosters you have, the more crowing that will go on because of each one communicating with his girls and controlling territory amongst the other roosters. When a challenge is made between roosters, they will generally face off in a sparring position. They will face each other with their heads dropped and their neck feathers raised. If neither rooster backs down from the challenge, then they will lunge at each other trying to kick and spur one another. This can go on until one or the other rooster backs down or the death of one of the roosters is the ultimate result, but this is rare; usually, one of the roosters will back down. The winning rooster is now higher on the pecking order than the losing rooster. The winning rooster will stand tall and let out a series of loud crows to announce that he has control over that area and to call his girls back to him.

The hens' challenges are generally seen at the feed bowl. Although you can have fighting between the hens, it is usually pecking at one another at the feeding area. It is fairly easy to determine who your lead birds are. If you have more than one rooster, take a handful of scratch out to the yard and throw it on the ground. Your lead rooster will go to the scratch and cluck wildly to attract the girls. The other roosters will stay back. The same is true for the hens. Take a bowl of treats out to the yard and place it on the ground. The hens will all crowd around the bowl. The lesser hens will generally stay back out of the way. Of the higher order hens, you will see one or two pecking at the others to get them out of the way. These are generally your top hens.

Pecking order is just that. If you were able to lay out a long line of feed, with one end being the best feed and the other being the worst feed, and watch the chickens as they eat, you would see that the lead hen would be at the front of the line at the best feed. The others would peck and push each other to find their rightful spot within the order all the way down to the bird with the least control. The society of poultry is a very complex and ever-changing one and within a large, diverse flock, is nearly impossible to keep up with. There are always challenges but most go unnoticed on a daily basis.

A rooster will also challenge you for dominance. This is the only reason that a rooster will try to attack you. Early on, while your birds are at the chick stage, your complete dominance over them must be established. As they get older, you need to continue to show your dominance. This can be done simply by interacting on a daily basis with your flock, especially with the roosters. In the evening, when they are going to roost, go in and pet each one, even hold them as much as possible. When they are in the yard and two youngsters face off in a sparring pose, stand between them to break them up. As your roosters get to the adult stage, promptly break up any challenges between birds. Hold your roosters as much as possible. Stand in their way when they are trying to get to a hen. All these things will show the rooster that you are dominant over him. This dominance can never be given up or you will end up with an aggressive rooster. But even aggressive roosters can usually be broken, depending on how long they have been allowed to be the dominant one in the flock. This is your flock and you need to be in control of it at all times. Granted, you can end up with a very stubborn rooster that continues to be mean, and for those the soup pot is usually the best option. Out of thousands of birds and hundreds of roosters, we have only ever had to send a couple to the stew pot. There is a pecking order in every flock, no matter how big or how small, and it is up to you to remain at the top.

Did you know?
Chickens can recognize and distinguish up to one hundred members of their flock.

BRINGING IN NEW BIRDS

Whether you are just starting out building your new flock or if you have a well-established flock, there will be times when you need or want to bring in new birds.

This would either be to expand your flock or to bring in new bloodlines to improve your existing flock. This practice has certain implications that go along with it.

Some people run what is called a closed flock. No new birds come in to their flock beyond what they hatch themselves. This is a fine practice and a fairly safe one as far as your birds are concerned, but it limits your breeding ability, as you do not have new bloodlines to further perfect the lines that you have. This is a common practice with larger breeders. After they have acquired the number of top quality birds that they need to perfect and propagate their flock, they then close their doors to any new birds coming in to protect their flock from outside disease contamination.

For people just starting out with a new flock or trying to improve the one they have, a closed flock is not always possible. New birds will have to be brought in. This opens the door for problems that people just do not realize are out there. You have to be prepared and willing to work with new birds prior to actually letting them enter your flock.

The first thing you will want to do when bringing any new birds onto your property is to quarantine them. A miraculous thing happens with chickens when they are moved: it never seems to fail that they get sick. But why is this? You go to a reputable person's farm and pick out one or two beautiful looking birds, bring them back to your farm only to have them be sick or even die. You call up the breeder you got the birds from and tell him, in no uncertain terms, what crappy birds he has and that you now have sick or dying birds of your own and that it is all his fault. Yes, there are some unscrupulous poultry owners out there that will knowingly sell sick birds, but they really are few and far between. They do not want a bad reputation for selling bad birds. Word spreads quickly in the poultry world.

Take a deep breath!

Those birds that you just got may have been perfectly healthy while on the other guy's farm. So why are they sick now? Chickens become immune to their surroundings. As they grow from chicks into adult birds, their bodies learn to fight off whatever diseases may be present in their surroundings. If you change these surroundings, then they are no longer immune. Chickens also have the ability to suppress diseases in their bodies. Chickens can be carriers of many diseases and it is stress that will make them come out. As a personal example, one day we purchased a set of beautiful, heritage Delawares from a gentleman. We had seen these birds a few times and knew that they were well taken care of. They were big, meaty birds;

the hens were laying well; no bugs; no visible problems whatsoever. The gentleman was going on an extended vacation and wanted his birds to go to a good home.

So, we went and picked up the birds and brought them back to our farm. I placed them in a separate pen and they all seemed very happy out there pecking and eating, taking dust baths and all. Within forty-eight hours, all of them were dead. No signs or symptoms and none of our regular birds ever had a problem after that. His birds were immune to their surroundings and by bringing them here there was just enough stress to cause something lethal to come out in their systems that proved rapidly fatal.

Even within your own flock, you may think that you have the healthiest birds around, and one day a predator of some type tries to get into your flock. The next day or in a couple of days, a few of your birds are now sick. You blame it on the predator—which, in part, is true. The bird already had the disease; it was the stress of the predator that brought it out.

The same holds true for moving birds. Moving a bird from familiar surroundings into unfamiliar surroundings is very traumatic to a chicken, especially if you just throw them in with other birds. They don't know where they are or why they were put there. They are being pecked at mercilessly. Nothing is the same and so they get incredibly stressed. Stress brings out disease in chickens much like it does in humans. If a human is stressed, they have a more difficult time fighting off disease and, therefore, seem to be constantly sick.

So what are you to do? Either you can keep a closed flock and be happy with the birds that you have, or you can deal with the possibility of sick and or dying birds. But it really is not as bad as it sounds. You have to take simple precautions to make the transition from one farm to another a more pleasurable experience.

It is up to you to make sure the bird or birds that you will be getting appear healthy. Buy birds from a reputable breeder or farmer. Running down to your nearest livestock auction and buying a bunch of birds that you know nothing about is not a wise way to increase the size of your flock. Make the transition from one place to another as quick and stress-free as possible. You are not going to want to have to travel hundreds of miles in scorching heat to pick up birds and bring them all the way back in cramped cages in the back of a hot pickup truck. Well, maybe you do, but you will not have healthy, happy birds when you get home. Do not just throw new birds into an established pen of chickens and let them figure it out with all the pecking and fighting. You are going to end up with a bunch of sick or dead chickens and then you will want to blame the person you got the chickens from when in fact it was your own fault.

To have the best experience with any new birds, it is essential that you make the process as smooth as possible. Check out any potential birds closely. Check for mites and lice in the feathers. Check for overall body condition; is the bird's weight correct for the age and size? Check the eyes and nose for any discharge. Ask questions about normal feeding routines. Look around at the chicken poop on the ground. Does it look normal, or is there a bunch of runny poop or discolored poop?

Once you have established that they are birds worthy of your purchase, then make the trip back to your home as painless and stress-free as possible. Place the chickens in large enough, suitable cages so that they do have some freedom of movement. If you have a fairly long distance to travel, place pieces of fruit in the cages so that the birds do not get dehydrated during the trip. Have the cages inside with you in air conditioning if it is going to be hot, or lightly covered in the bed of a truck if it is cooler weather. Protect them from the elements. The bed of a truck can get extremely hot even on cooler days, even if you are going down the road at 60 miles per hour. Keep the cages elevated to provide space so that the birds do not have to sit or stand in their poop or on hot surfaces. And make the trip as short as possible. The least amount of time the chickens are stuck in the cages, the happier they will be.

Once you get the birds back to your place, they need to be put into quarantine. It doesn't matter whom you got the chickens from, they still need to be separated from your regular flock. Your quarantine area needs to be as far away from your normal pens as possible. The farther the better, as many diseases are airborne. This quarantine pen needs to be of suitable size to allow the birds natural freedom of movement. It needs to be protected from the elements and provide the birds with adequate food, water, nesting and roosting space.

Your new birds need to remain in this quarantine pen for up to thirty days. Yes, that is correct, up to thirty days. This will give time for any problems to arise. Many diseases have seven- to fourteen-day incubation periods before they show symptoms. This gives you the opportunity to observe the birds and to determine if there are going to be any problems and also to correct the problems that do come up.

After the quarantine period, if all is well, then you can transition your new birds into your established flock. Properly transitioning birds helps alleviate unnecessary stress on the birds as well. Now all this is not to say that you still won't have some problems, but you will not have nearly the trouble that you could have.

You know you're addicted to chickens when . . . you take more pictures of your chickens than of your own children.

TRANSITIONING CHICKS AND CHICKENS

Almost everyone who comes to visit us here on the farm asks us how we get our different-aged poultry to all live in harmony in one pen. Some days we wonder the same thing, but what works for us is this transition period before they are actually placed together.

If you have ever put two chickens together or been around someone who has, you know that they will begin fighting almost instantly. This can be very frightening for the poultry owner who is simply trying to expand his flock. But there is an easier way to maintain peace and order in the barnyard.

The use of a smaller pen adjacent to or within your larger pen is the easiest method we have found for transitioning new birds into the flock. Let's cover a few basics first. If you have chicks that you want to transition into your flock, they should be at least six weeks old and fully feathered before being placed in an outside run; any younger than this and they risk illness due to exposure. As mentioned above, if you need to transition older birds into your established flock, you should have had them in a quarantine area for the past thirty days to watch for any illness that might come up.

With that said, if you have one main pen that your normal flock is in and you want to transition the new birds into it, a second pen is needed. This can be a simple temporary structure built within or adjacent to your existing pen. It needs to have adequate shelter along with food and water. This will be the new birds' home for a few weeks. At least one side needs to have one-inch poultry netting all the way to the ground and preferably buried at least six inches deep. This is what will separate your main flock from the new birds.

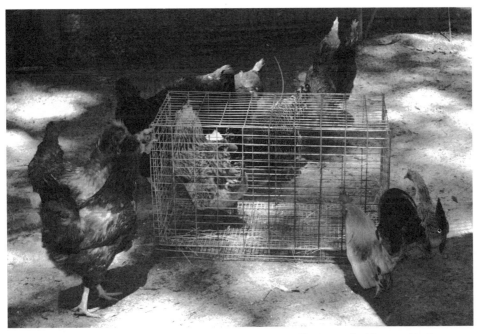

Using a smaller pen adjacent to or within your larger pen is an effective method for transitioning new birds into the flock.

Place your new birds inside this area. Your established flock will come and check out the new birds and both sides will jump and claw at the fence trying to get at each other. This is the reason for the one-inch mesh separation. Generally, after about an hour, your flock will go about their business with only minor altercations at the fence line. Over a few days, all the birds will pretty much ignore each other unless you have roosters in each pen.

Within the first week, a pecking order will be established through vocalization, scent and also pecking through the fencing. After two weeks, you can open the pens up to each other and let the birds mingle. Do this at a time when you can be present. Your established flock will want to get into the transition pen to see what they have been missing out on. The new birds will come out into your main pen and stay in a group to themselves. There may be some minor pecking, but just break them up and they will soon go on their way. When nighttime comes, allow your new birds to return to their area for roosting at least for the first couple of nights.

As the birds all start getting used to each other, you can then work on getting the new birds to roost with the older birds. Take each new bird off their roost and place them separate from the established birds upon the regular roost. This way they all wake up together and they start to get used to sleeping together. On the second or third night, as nighttime comes and the birds want to roost, close the door to the transition pen. Slowly work the new birds towards the roosting area. The darker it gets, the more easily they can be led. Let them pick their place along the roost. This may have to be done a couple nights in a row, but they will soon learn where they need to go. After this time, you can then remove the transition pen or leave it for future use. This is how we transition all our birds and it works for us every time.

Did you know?
Chickens have over twenty-four known different vocalizations and each breed has its own tonal range of vocalization.

CHANGING A FLOCK'S DYNAMICS

Your chickens live in a hierarchical society, but sometimes the power that is established in the lead birds goes to their head. There is unnecessary pecking and bullying going on within the flock, and on occasion a bird needs to be brought back to reality and taken down a peg or two.

Every time that you bring in a new bird or take one away, you change your flock's dynamics. This can be because you are bringing in a new bloodline or because you had a death in the flock. Whatever the case, the dynamics of your flock changes. This same theory can be used to change the attitude of a bird.

Let's say that you have a lead hen that has taken to bullying the younger hens for no apparent reason. This is going way beyond the normal pecking order stuff. She is chasing them down and pecking at them out of sheer orneriness. You are tired of having to break up the squabbling and are about ready to get rid of the mean old hen. Take a deep breath.

You need to change her status in the pecking order. You need to forcibly change your flock's dynamics. But how is this accomplished when the chickens establish their own pecking order? It is quite simple. Have you ever had to put your child in a time out because of bad behavior? The same holds true with chickens. They just need a time out. Any time you can take a bird temporarily out of the flock, you change your flock's dynamics. A different bird will take their place amongst the pecking order.

We have found that placing a bird that needs an attitude adjustment into a separate pen away from the main flock for three to five days will generally do the trick. Sometimes it takes a bit longer, but that would be the rare occasion. We have also found that we can take an ornery hen and put her in a smaller cage and place her into one of the breeding pens. After a couple of hours, we can release her into the breeding pen and now she is the low chicken in the pecking order. We can leave her in the breeding pen for a day or two and then return her to the main flock and her attitude is miraculously changed.

Because of what we do, we have a pretty open flock around here. Birds come and go. So with people that we know and birds that we are very familiar with, birds will sometimes be brought here for attitude adjustments. Most of our birds are very good-natured and very easily handled. We also have a large flock of birds. If a friend brings us a bird that they have that is being ornery to their other birds, we will place it in a cage in the middle of our main poultry yard. Our birds will check it out for a few minutes and be on their way. We can then release the bird into our flock and now it is at the bottom of the pecking order. There may be a bit of chest bumping or a bit of a squabble at first, but they soon settle into pecking and scratching along with the rest of them. After a couple of hours of this, the bird can be returned to its original flock, usually without further incidents.

Simply by the use of forced change in your flock's dynamics, you can manipulate the pecking order, and in exchange have a more peaceful and enjoyable poultry experience.

POULTRY PERSONALITY

As more and more people are coming to realize that chickens make great pets, the people around them are wondering if they have just gone plum crazy or if there really is something behind this chicken craze. How could anyone ever make a pet of a stupid creature like a chicken? After all, they have a tiny brain, cannot be trained, and all they do is eat, poop and squawk. The truth is, every chicken has its own personality. It is no different from your typical dog or cat.

Throughout the years, flocks of chickens were left to their own devices to peck and scratch around the barnyard and were relegated to laying eggs and providing the occasional Sunday meal. Not much thought went into them. If a hen came up to you when you went outside, it was because she wanted something to eat. If a rooster attacked you, it was because it was mean and quickly headed for the soup pot. But a chicken really is a very complex creature that we are just now starting to understand. No, you may not be able to teach it to roll over, sit up or bark on command, but then I have never seen a cat that would do that either. Or a rabbit, or a snake or a goldfish. But spend a day with a chicken and you will be amazed at what you will discover and learn.

The time we spend watching our flock, we call "Chicken TV." There are fights and squabbles, love interests and errant matings. There is pouting and dejection as well as joy and celebration. There are outside forces at work as well as criminal behavior. There are dastardly villains right along with the heroes. It has all the makings of the best soap operas. But watch your chickens closely and you will see that each one has its own personality. And it is each of these individual personalities that make up the dynamics of your flock.

Within our flock we have DeDe and FiFi, who will follow you around the yard wanting to be held. There is Squatter, who wants to be tickled. Bruno runs the yard. He is the hero and comes to the aid of all the girls. Miss Prissy is the mothering one and Dell is the alarmist. There are the shy ones and the bold ones, the quiet ones and the ones that chatter all the time. Buttercup talks like the stereotypical trucker. Each and every one of our birds has a different personality.

Each individual personality is developed from a very young age, and just like in humans, will change as they get older. You can see the differences in a batch of day-old chicks fresh from the incubator. Most will run from you because they do not yet understand, but sit quietly with them and put your hand out at their level and soon one or two will come check you out. Rub their chest a bit and soon each time you stick your hand in their brooder, those couple will come running to you to be held. As they get older they will start determining their pecking order and you will have the dominant ones and the submissive ones. A bit older and they become teenagers and all bets are off. Attitude goes through the roof, but luckily it does not last as long as with human children. They settle back down and get into their daily routine. Groups and subgroups are formed and personalities are really formed and become visually apparent. The hens will start laying and the roosters will be roosters. But it is up to you to help shape their true personality.

At this moment as I write this, Mille is sitting beside me preening herself. She has been in the house for a week or so because she decided to come down with fowl pox. At night she is in a cage but during the day she has the run of the house. She has taken a real liking to me since she has been inside, and now if I leave her sight she will chatter up a storm until I come back or she comes to find me. She will usually sit right between my arms as I am typing. At night, if I am watching TV, she wants to sit on my belly and be petted. Mille is a little Serama hen. We also have a large Rhode Island Red rooster named Little Rock in the house. He is recovering from a respiratory infection. He is young but he is still quite large for his age. Mille has kind of taken a liking to him and he to her. We let him out during the day as well and they both come and go as they please. The front door is open so they can get outside to do their chicken thing. Mille, being the older of the two, has started showing Little Rock the finer points of grubbing in the dirt. She will cluck and scratch and then peck at the ground to show him where something is. Little Rock is big enough to squish Mille but he is very gentle around her. She tells him how it is and he listens.

Little Rock prefers to be outside and Mille prefers to be inside, especially if I am in the office working, but she will get up every so often and go outside to look for him, then she will come back in and get right back up on the desk so she can be beside me. She will be in for a shock when her fowl pox clears up completely and she has to go back to the coop with her other like kind, but she will adjust. Her best friend Abby is waiting for her return.

Chickens, contrary to popular belief, are not stupid creatures. Yes, they are creatures of habit, but they are far from stupid. They live in a very complex society made up of groups and subgroups. They have at least twenty-four different vocalizations and each breed has its own range of vocalization, yet they can all recognize each other by sight and sound. They can easily recognize up to one hundred other members of their flock. We have birds that are just as friendly as can be to most people, but then be very standoffish when other people come into the yard. Our birds know that when the blue bucket comes out that that is feed, and if the hot pink scoop comes out then it is treat time. They know that if they go over the perimeter fence, they will get in trouble and will sulk when we shoo them back in. The boys know that if they are misbehaving, they will be scolded and if they keep it up, then they are sent to their own pen. Some of the girls will come when they are called. No, they can't do a bunch of fancy tricks, but then again we have never tried to teach them.

The hens all have a favorite nest and will cluck loudly when someone is in there and they want in to lay an egg. And a broody hen is impossible. Her instincts kick in and it is like she has a twenty-one-day case of PMS. If you work with your chickens each day and learn from them and they from you, it is not long before you see their true personalities come out. This will help you with your daily dealings with them. You will know who can be touched and who can't, who is dominant and who is submissive. They are full of antics and are very enjoyable to just sit and watch. You will find that the ten minutes you intended to stay in their pen has soon turned into an hour or more. Chickens are very intelligent animals and deserve our attention to their needs and desires. They will reward you with experiences that you can tell all your friends who think you are crazy. They are really the crazy ones for not being able to experience the joy that raising chickens can bring.

Did you know?
The chicken is the closest living relative to the Tyrannosaurus rex.

AGGRESSIVE ROOSTERS

There is nothing worse than having to go into your chicken pens or walk out into your yard and be in fear of your rooster attacking you. A full-size rooster can do incredible damage to the human body to the point of even breaking bones. No one should ever have to put up with this or be in fear of his or her birds.

A rooster has an inherent need to be the leader of the flock. He protects his girls from anything that poses a threat. A rooster will fight anything that he sees as a challenge to his leadership.

Every flock is a hierarchical society. This means that there is a certain pecking order that has to be maintained. Generally, the rooster is at the top of the pecking order, followed by a lead hen and down from there to the lesser birds. If you have two or more roosters in your flock, they will be constantly vying for the top spot. This is normal behavior. What is not normal is when the rooster attacks you or anyone else who enters their domain. A rooster wants to dominate everything within its area and he will also want to dominate you.

People will write, phone or stop by our farm to tell me that they have the meanest rooster around and go on about how it always attacks them and I just nod my head in approval as I listen to their tales of woe. And then they look at me funny when I tell them that there is no such thing as a mean rooster. The rooster is merely doing what a rooster is supposed to do and what you have taught it to do. He is simply protecting what is his and what you have allowed him to have.

A rooster has an inherent need to be the leader of the flock and will fight anything that he sees as a challenge to his leadership.

You are the master of your flock and you have to be the dominant one at all times. It is up to you to establish that he is yours and the hens are yours and the pen is yours and even the ground he walks on is yours. It is up to you to be at the top of the pecking order and there are some simple ways to achieve this. The worst thing that you can do is to carry a stick or rake or what have you, when confronted by a rooster. He will only learn to fear it and not you.

The best way is for you to stand your ground against him. Roosters attack with their feet. They will beat you with their wings, but that is simply because they are trying to confuse you and strike you with their spurs and use their wings to get them into a position to do so.

There are two methods that we use on the farm to establish our dominance over the roosters. They vary greatly in methodology, but both achieve the same result.

The first is the method that we use most. Catch your rooster. If you have an overly aggressive rooster, this might be better done at night so that he can be placed in a smaller cage for easier handling. Leave him in the cage overnight and start this process first thing in the morning. Once you have ahold of your rooster, cradle him in your arm so that he is tight against your body, much like you would carry a football. With the arm that you are cradling him with, also use that hand to lightly restrain his feet so he can't kick and get away. Now with your other hand, take ahold of his comb and gently pull his head down to where his beak basically touches his chest and hold his head there. He will want to struggle to get free but you are not hurting anything but his pride. Talk softly to him and just walk around with him. Continue to hold his head in this position until he stops struggling. Slowly release his head. If he picks his head right up, take a hold of his comb again and resume the position. Continue to do this until he holds his head down on his own and does not raise it unless you raise it for him. This can take five minutes or it might take a couple of hours depending on how stubborn your rooster is.

When he is willing to hold his head down on his own once you place it there, release him onto the ground. He should walk away from you. You may have to do this a few days in a row, but you will establish that you are dominant over him and he should walk away from you every time you enter his territory. I have seen this done on fighting cocks and it works within ten minutes. Our roosters are hand-raised from chicks and so we rarely have a dominance issue, but when we do, this method works quickly.

Each family member may have to do this independently so that the rooster understands that you are all dominant over him. Even small children can do this with a large rooster; you may just have to help support the rooster's body weight.

The whole premise behind this method is that the head-down posture is submissive in the world of chickens. If you watch two roosters sparring, they will go head to head with their hackles raised. They will mirror each other's movements until one drops his head just a fraction of an inch. This gives the other rooster an advantage and he will strike. If a chicken gets cornered by another chicken, it will generally drop its head in submission. By holding your rooster's head down, you are forcing him into submission.

The second method is more old school. Stand your ground. As the rooster comes at you, do not turn away from him. Raise your arms and, using the inside of your foot, give him a good boot. Boot him hard enough to send him tumbling away. You are not out to hurt him, merely startle him. Try to aim as best you can for his chest or just to the side of his chest. Continue to stand your ground. He may come at you three or four times before he figures out that you are the top of the pecking order and you are not going to back down. He will eventually go away dejected. With a stubborn rooster, you may have to do this a few days in a row, but he will learn. He will also try to regain dominance every three months or so. Don't turn your back on a dominant rooster because, if they think they can take a shot at you, they will.

The same thing holds true with other family members or even friends. Just because your rooster respects your place in the pecking order doesn't mean that he will respect others. If he attacks your spouse, then have your spouse also boot him until he learns his place with them. As for the case with smaller children, you may have to do the booting for them, but do it with the child present.

This is not cruel. What is cruel is using a rake or stick, because he will never understand his place.

If either method fails, then it would be best to get rid of the rooster or we have some great chicken recipes.

Bathing and Grooming Your Chickens

Author's note: This section was kindly contributed by Cindy Kinard, APA-ABA Florida Youth Program Leader.

Chickens in a natural environment will take dust baths every day or at least every couple of days. This is done by them scratching a hole in the dirt, lying in it and throwing dirt over their bodies with their feet and wings. This helps them clean the excess oils from their feathers and also helps rid their bodies of parasites.

But with today's chickens being kept as pets and also being entered in poultry shows, it is necessary for us to give them a helping hand by providing them with a spa day. I know, ladies, you are thinking, "Spa day? I don't even get a spa day for myself!" But then again, we spoil our chickens. Plus, it is much cheaper. But, don't get me wrong, you ladies are worth it.

You never want to take a dirty bird to a show. Dirty birds don't win; they don't even get as far as champion row. The next time you are at a show or a fair, take the time to look at the birds. Really look at them. Stand back and see if you can see what the judge sees. Why did the judge place one bird over the other? In some larger shows, the placement can be very close and the difference between your bird being placed on champion row or left in the cage can come down to its being clean and well groomed.

By the same token, if you bring your birds into the house with any frequency, you are not going to want to have a bird that is all dirty running around on your carpets. Chickens sometimes just get nasty running around the yard and need some general maintenance. Bathing and grooming also helps your bird to remain healthier by keeping its feathers free of foreign materials and nasty little parasites that can cause all sorts of problems.

Bathing a chicken is not some big secret that only the top breeders know. Not only is it easy to do, your bird will actually enjoy its bath. In the case of chickens that are going to be placed in a show, bathing is done three to five days prior to the show date. This allows time for the bird to dry completely and to work some of the natural oils back into the feathers. Grooming starts at bath time with the trimming of the nails and the beak, and the final, last-minute touch-ups take place after your bird is in the show cage and just prior to the judging. If you do not have show birds, then this time line is not so critical and bathing can be done when it is most convenient for you.

Gather your supplies and place them within easy reach. This list is simple and basic. Of course, you can be more elaborate if you wish, but I have found that the basic method works well. List: 3 washtubs, shampoo, vinegar, hair conditioner, towels, an old toothbrush, dog nail clippers, a blow dryer, a small sponge, and a clean carrier or cage for drying. Some of the extras that you might want to have on hand are: an emery board for making smooth nail and beak trims and blood stop in case you cut the nail too short.

Start with the three tubs of warm water. The tubs should be large enough to give you plenty of water and room for a good bath. Don't try and bathe a large fowl bird in a small plastic bucket. It just won't work. Set up your washtubs in a suitable location that can get wet should the bird start flapping its wings, such as a laundry room, mud room or, if the weather is warm enough, out on the back porch. Fill the tubs with enough water to easily cover the back of the bird. A chicken's body temperature is normally from 104°F (41°C) to 107°F (42°C). Your wash water should be slightly warmer than this like around 110°F (43°C).

The first tub is used for the bath while the second and third tubs are used for rinsing. Put about ¼ cup vinegar for every gallon of water in the second tub. Add about 1 tablespoon of hair conditioner for every gallon of water in the third tub. Holding your bird with its breast resting in the palm of one hand and your other hand over its back to hold its wings from flapping, lower your bird into the first tub, allowing a few seconds for the bird to realize what is happening. Most birds will relax and some will even go to sleep during their bath. This really happens, so keep a watch on your bird so their head does not go under the water.

Put some shampoo on the dirtiest part of your bird first. This is usually the vent area, the legs and the feet. While these parts are soaking, shampoo the rest of the bird slowly and gently, being careful that you do not damage the feathers by rubbing them backwards. Work the shampoo all the way down to the skin. Take your time and do a good job. Rinse as much of the shampoo out of the feathers as you can in the first tub and then place the bird in the second tub with the vinegar. Rinse, rinse, rinse. The vinegar will help but you may need to rinse again. Make sure there is no shampoo left in the feathers because this makes for a mess when it dries. (Think of drying your hair with shampoo left in.) Do your final rinse in the third tub.

Next, lay the bird down on a towel and wrap the bird in the towel, leaving the head out of one end and the feet sticking out of the other. (Think egg roll.) This will help dry the bird and keep it still so you can clean its head, legs and feet. While it is wrapped, you wash the face, wattles and comb with a damp sponge. Trim the top beak so that it is even with the bottom and use the emery board to smooth the edges (details of this are covered in the Trimming Beaks section found later in this chapter).

After a good bath and face washing, it's time for a good pedicure. Using the old toothbrush, give the legs, feet and toenails a scrubbing. Make sure you remove any old dirt that is under the nails, as judges do notice. You may need to use soap to do a complete job. Rinse them well under running water. Now clip the nails. It is easier to clip nails once they have softened in the water and are clean. Be careful not to trim too short as they have a vein that runs down through the toe and into the toenail. Clipping into this vein will cause them to bleed. This is usually not serious, but it can bleed, and remember that you have just bathed your bird and you do not want to get blood on the feathers. The vein is fairly easy to see in birds with white toenails, but on dark ones, you may have to look under the nail to see where the vein stops. If you cannot see the vein, clip small portions of the nail at a time, checking after each clip. If the nail does bleed, use the blood stop and a cotton ball to control the bleeding. (Nail trimming will be discussed in better detail later in this chapter.)

Now you can put your bird in the clean cage or carrier to dry. While all birds can be air dried in a warm environment (not in the sun or in front of a heater), loose-feathered birds such as Cochins, Silkies and Orpingtons will benefit from drying with a blow dryer, while tightly feathered birds such as Old English and Modern

Game do better drying on their own. If you are going to blow dry your birds, do so with the dryer set on a low setting so as to not burn the feathers or the bird.

Allow your bird time to fully dry and fluff its feathers before returning it to the yard. In the case of birds going to show, you will want to keep them in cages off the dirt until you are ready to transport them. With birds going to the show, transport them in a carrier with smooth sides to protect the feathers that you worked so hard to clean. Once at the show you will need a grooming box (a small tackle box or five-gallon bucket) stocked with your grooming supplies. Here is the basic list: baby oil to put a shine on the shank, feet, comb, beak and wattles of the bird; blood stop in case you need to stop bleeding from anything; an old toothbrush for cleaning toenails; wet wipes; antibiotic ointment to use on combs or wattles that are scratched and bleeding; and a silk cloth for wiping feathers and putting a shine on the bird. These things are not necessary for pet or yard birds unless you really want to spoil them.

After cooping in your bird at the show, and about thirty minutes or so before judging begins, do your last-minute grooming. Start with the feet and legs by wiping them with the baby wipes and then putting on a little baby oil for shine. Check the vent area for any poop that might be on the feathers, and clean them with the wipes. Check the head and decide if you want to put baby oil on the comb and wattles or if antibiotic ointment would be better. Either way, rub it in well so it produces a nice shine. Now use the silk cloth and rub your bird from head to tail several times; twenty is good but fifty is better. The bird will enjoy this and you will see those feathers begin to really shine. Gently place the bird back into the cage so as not to disturb those beautiful feathers. As you attend more shows, you will see other breeders groom their birds and you will learn more techniques and see other products used on birds. You can learn much from other breeders and you may want to try some of what you learn. Remember, never do anything to your bird or use anything on your bird that might harm it in any way.

Now you can walk away knowing you have done a beautiful job and that your bird looks its best.

CLIPPING WINGS

Depending on how you have your chickens housed, you might find it necessary to clip their wings to keep them from escaping over a fence. Clipping wings is a very easy process that does not harm the bird if done properly. You will only want to clip one wing of the bird. If you clip both wings, they will learn that if they flap their wings faster they can still fly. By only clipping one wing, it throws their body off balance enough to discourage flying. Once clipped, the wing feathers will grow back in about four months, so the procedure will have to be repeated. When we clip wings, we clip all the left wings one time and then alternate and clip all the right wings the next. This discourages the birds from learning to fly even though they have one clipped wing.

To clip a chicken's wing, you simply need a sharp pair of scissors. Hold the bird cradled in your arm like you would a football. Have the bird's head pointed towards your elbow. With the hand that you are holding the bird, spread out the wing to be clipped.

You will only be clipping the primary flight feathers, which are the ten long feathers to the outside of the wing. These can be distinguished by looking at the feathers of the wing. Look at the long feathers. At about the middle of the wing there is one short feather; this is called the axial feather. All the long feathers to the outside of this feather are the primary flight feathers. All the feathers to the inside of this feather are the secondary feathers.

If done properly, clipping wings is very easy and doesn't harm the bird.

Never cut the feathers higher than the covert feathers; you risk hitting the viens in each feather which could result to your bird bleading to death.

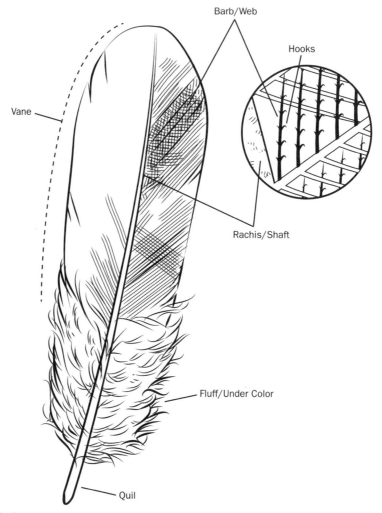

Parts of a feather

Looking at the primary flight feathers, you will see a second row of shorter feathers; these are called the primary coverts. On a large fowl bird you will be cutting off the primary flight feathers approximately one inch below the primary coverts. Position your scissors so that you are cutting from the first feather past the axial feather and out to the outside edge of the wing.

You should be able to cut all the primary flight feathers in one smooth cut. That is all there is to it.

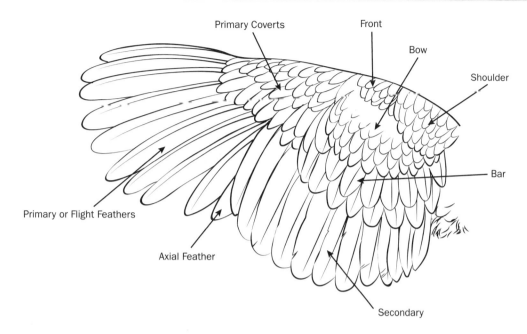

Parts of a wing

Never cut the feathers higher than the covert feathers as you risk hitting the veins that are inside each feather—this could possibly result in your bird bleeding to death. Never cut the length of the wing as you will do permanent damage to the wing and also cause the bird great pain and distress that could lead to the death of the bird. Only ever cut the flight feathers as described above and you and your birds will be happier for it.

TRIMMING TOENAILS

Yes, even your hens need to be divas once in a while. Birds that are kept in small coops, in elevated cages or not allowed to free range, will on occasion need to have their nails trimmed. This should be part of your monthly regimen of general maintenance. If nails are left to grow, they can cause great pain and discomfort to the bird and can cause lasting injury to the foot.

Trimming a chicken's toenails is really no different than trimming any other animal's nails. Start by washing and scrubbing the bird's nails with warm soapy water and an old toothbrush. This helps remove any dirt, feces and old scaly tissue from the foot area. Dry the foot. Looking at the side of the nail, you should be able to see a faint difference in color along the length of the nail. The darker center line originating at the toe is the quick. This is where blood is in the nail. You do not want to cut within this area or the nail will bleed and you will cause unnecessary pain to the bird.

Any type of nail clipper will work for trimming their nails, but we prefer the guillotine-type clippers as we find they are much easier to clip the tough nails with.

Holding the chicken cradled in your arm with its head facing forward, use that hand to spread the toes apart. Using the clippers, clip short sections of nail off at a time, being careful to not cut into the quick.

Move from nail to nail on each foot until complete. If by chance you should cut into the quick and the nail starts to bleed, use blood stop powder, flour or any other coagulant to help stop the bleeding.

In the case of extremely long, overgrown nails, it may be necessary to clip the nails in intervals over a few weeks to be able to get them back to a normal length. By working in intervals, this allows the quick to recede to where you can cut the nails shorter. Once the nails are to the desired length, you can file them if you wish, or by natural scratching the nails will file themselves back into a pointed shape.

And yes, some people go as far as to paint their birds' nails. This is fine, but beware that flashy colors can cause the other birds to want to pick at them and may cause trauma and injury to the bird.

Did you know?

An experienced rooster will not try to mate with a young pullet until it is about two weeks away from laying its first egg. This is a good indicator of when young hens will start to lay. Other things to watch for would be reddening of the comb and wattles, and the vent becoming oval instead of round.

TRIMMING BEAKS

There is a huge difference between trimming a bird's beak and actually debeaking them. We do not believe in debeaking chickens, and if they are kept under humane conditions, there is no reason to ever debeak a chicken. Debeaking is primarily done in the large commercial egg-producing plants to keep the chickens from cannibalizing each other. If you are experiencing a cannibalism problem, then you have other issues that need to be addressed such as overcrowding, boredom or nutritional deficiencies.

Trimming beaks is just a matter of keeping the upper beak in line with the lower beak so that an overhanging beak does not hinder eating and pecking. Trimming also helps prevent the beak from cracking and breaking. When the top beak overhangs the lower beak, it needs to be trimmed.

Trimming the beak is a quick, simple procedure that can be done with ordinary nail clippers and a nail file. This procedure is possible with one person but much easier with two. As one person holds the bird, gently secure the head of the bird in your hand and with your fingers hold the beak closed. Take the clippers and hold them just about perpendicular to the bottom beak and trim off any excess of the top beak so that they now match in length.

Use the nail file to gently file the beak into a more natural pointed shape that matches with the bottom beak. As you can see in the picture below, the upper and lower beaks now match in length, making it easier for the bird to eat and also lessening the risk of cracking or breaking of the top beak. It is quick, simple and painless for the bird.

If you should have a bird that has cracked its beak lengthwise, depending on how severely cracked, it is possible to carefully glue it together using super glue. Obviously, you would want to be extremely careful doing this as you could inadvertently glue the bird's beak shut. Once it is glued together, you can put a thin coat of epoxy, such as J-B Weld, over the top of the beak, being careful not to cover the nostrils, then gently file the beak back into form. The beak will slowly grow out and can then be trimmed as

Trimming keeps the upper beak in line with the lower beak so an overhanging beak doesn't hinder eating and pecking.

Trimming beaks is much different from debeaking, which takes place in large commercial egg-producing plants.

needed until the crack is gone. Placing the bird in a cage and giving wet mash will help keep the bird from having to peck so hard for food.

If the beak should happen to crack or break crosswise, there is not much you can do about it. Depending on the severity of the break, you can certainly try the super glue and epoxy method, but it is doubtful that it would hold. There have been cases of people fashioning a repair out of fiberglass, but I have never personally seen it done. Otherwise, you could leave it as is, and it would be as if you had debeaked the bird. It will learn to eat this way, although it will be difficult.

Use a nail file to shape the beak into a more natural, pointed shape that matches the bottom beak.

SPURS

Roosters have sharp, pointed protrusions on the insides of their legs called spurs that can cause serious pain and injury to you or another animal. Hens can even grow spurs, though it is not common. The Sumatra roosters have more than one set of spurs.

The roosters use spurs as weapons during a fight. The rooster will jump up, flap its wings, and slap its feet together hoping to inflict pain and injury with its spurs. If a rooster jumps at you and by chance hits you with his spurs, he can cause very serious injury.

If a rooster's spurs get too long, they can affect the way he walks and can also cause harm to the hens when he tries to mate. The spurs can be kept short through a simple procedure that removes the outer sheath and does not hurt the rooster.

If the spurs are exceptionally long, hold the rooster in one arm, kind of like you would hold a football. Taking hold of the spur with your free hand, and using a twisting motion, twist it back and forth until you feel it pop loose. This may take a

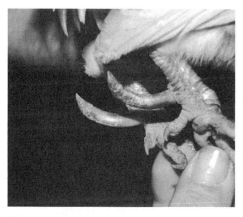

Roosters have sharp, pointed protrusions on the insides of their legs called spurs that can cause serious pain and injury to you or another animal.

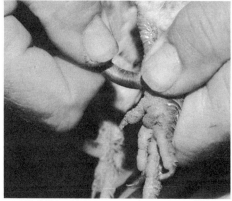

Spurs can be kept short through a simple procedure that removes the outer sheath and does not hurt the rooster.

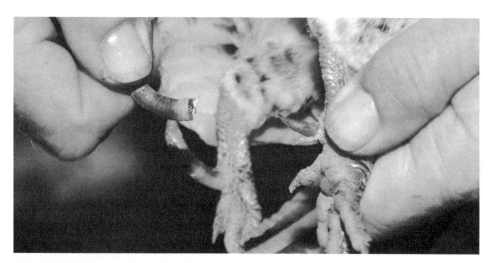

Taking hold of the spur with your free hand, and using a twisting motion, twist it back and forth until you feel it pop loose. This may take a bit of force, but do not use too much force as you will injure your rooster. The under spur may bleed a bit, but it will soon stop.

bit of force, but do not use too much force as you will injure your rooster. Simply slide the spur sheath off of the softer spur bone underneath, called the calcar bone, and move on to the other spur. The under spur may bleed a bit, but it will soon stop, or you can put a blood stop powder on it if you choose.

If the spurs are shorter, you may need to use a pair of pliers to be able to get a firm enough grip on the spur sheath. Again, you do not want to rock the spur back and forth, but instead use a twisting motion.

The under spur will be somewhat softer than the spur sheath. This under spur becomes the new spur; it will harden in a few days and a new sheath will form. If the spur is still too long, then in a couple of months you can again remove the spur sheath to reveal an even shorter under spur.

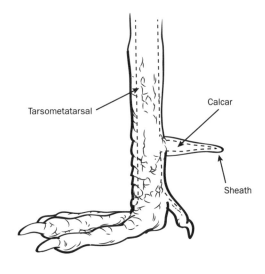

Spur of a rooster

Spurs are like fingernails. They will continue to grow back and therefore will need to be removed periodically. A short-spurred rooster is much safer than a long-spurred one. You are not hurting your rooster by removing his spur sheaths, and you will be doing him and you a great favor.

There are other procedures to remove the spur, but they are not recommended. When the chicks are very young, a trained veterinarian can remove spurs through a procedure using cauterization. This will keep the spurs from growing at all. Spurs on older roosters can also be removed—again, by a trained veterinarian—but this would take invasive surgery. Because the calcar bone is actually a protrusion from the leg bone, or tarsometatarsal bone, this requires the rooster to be put under anesthetic and the bone cut away to remove the spur. The skin must then be sewn over the surgical area and a period of recovery afterwards.

There is a school of thought out there to cut the length of the spur with a pair of suitable cutters such as wire cutters from the tool chest. This is highly discouraged. Not only does this cause great pain to the rooster as well as extensive bleeding, it also can bring on severe infection to the leg bone and surrounding area. The rooster would likely be maimed or disabled by this procedure.

The least intrusive process involves simply filing down the end of the spur. This does not necessarily shorten the spur to any extent but it does make it more rounded and, therefore, less likely to penetrate skin should the rooster become aggressive. This can be done with a hand file or a Dremel rotary tool.

A rooster's spurs are formidable weapons, and attention should be paid to them in your regular poultry maintenance program. By using an approved method of care, your rooster will be healthier, happier and less likely to inflict injury. It will also give you and your rooster a bit of bonding time, which is always a good thing.

Molting

You walk out to the hen house only to find that it looks like your girls had an industrial-size pillow fight. What is going on? Why, they are molting, of course, or at least that is what we are hoping.

Molting is the act of shedding old feathers to be replaced with a new heavier coat of feathers in preparation for winter's cold. Molting is much like a dog shedding, only it happens during different seasons. A dog will shed most of its fur in preparation for the warm months of summer, whereas a chicken will shed its feathers in preparation for winter. This is not to say that they can't molt at other times of the year, but your heaviest and most consistent molts will be in the fall.

All chickens molt. It is a natural occurrence and the hens molt much more severely than the roosters. Depending on the age of your chickens, this will determine if they go through a heavy molt or just a light molt. It will also depend on weather conditions and diet.

Generally, anywhere between September and November, your hens will lose masses of feathers in a very short period of time. It is possible for them to be nearly naked. But at the same time new feather shafts are coming in. They will very much look like a porcupine.

Chickens lose feathers all year long. One will fall out and be replaced by another. But molting is a near complete loss of feathers to be replaced with a heavy layer of down and all new feathers. A young pullet who was hatched in the spring will usually only go through a light molt her first year. This is because, as she has feathered out over the year, her feathers continue to get thicker and thicker. Come fall, she has most of the necessary feather covering to keep her warm through the winter. She will go through a light molt to add the down that is needed. A hen that was hatched the previous fall, or a hen that is into her second year, will molt much more severely.

Roosters also molt, but it is much less dramatic. They will lose a very small portion of their body feathers, but their tail feathers will be shed. This is true in most breeds of chickens other than the long-tailed breeds such as the Onagadori, which will only shed their tails every three to four years.

Molting also has a role in the reproduction cycle. Hens, as nature intended, will go broody in the early spring. The weather is still very cool. This added layer of feathers helps to keep the eggs warm, as well as the newly-hatched chicks. She will also pull some of the feathers off of her belly area to help line the nest. If a hen were to molt just prior to sitting on her clutch of eggs, she would likely freeze to death from exposure. So nature dictates that the hen molts in the fall so that she will be ready come spring.

During a molt, the hen will slow down or stop her laying. Her body is taking all the extra nutrients to produce new feathers. She requires higher levels of protein during this time. Feed supplements or a change in feed can help her through her molt. By adding such things as small kibble kitten food, black oil sunflower seed or any other high-protein food to the diet, the hen will grow feathers in at a faster rate and also possibly not stop laying. If kept on its standard diet, a hen will take up to two months to complete her molt. By adding extra protein to the diet you can generally cut this time in half.

There are other forces at hand that can cause your birds to lose lots of feathers, though generally not as bad as a molt. Wild animals can get into your pen and pull out feathers. Rats will chew off tail feathers. Feather mites can cause feathers to fall out, as can a poor diet. If you notice a large amount of feathers in a pen, these would be the first things to check for, especially if it is not fall.

Molting is a natural process and a much needed one. There is very little that you can do about it. There are ways to force a molt, but this is usually left to the commercial egg producers. To help your girls through their molt, you can add protein to their diet and make sure they have plenty of shade to get out of the sun. Beyond that, know that they will soon recover and look better than ever.

You know you're addicted to chickens when . . . your best Sunday shoes have chicken poop on them.

Why Have They Stopped Laying?

Many times throughout the year, we get calls, e-mails or personal visits from folks wanting to know why their birds have stopped laying. My first answer to them is stress. Yes, stress. But just like in our own human lives, stress can come in many forms. In the poultry world, stress can come from illness, a move, a change in your flock's dynamics, seasonal changes, predators—even from the weather. It will depend greatly on the individual bird's ability to deal with the stress as to whether or not it slows down on laying or completely stops.

But there are also things other than stress that make a hen stop laying. Some breeds of poultry are just not good layers. Take for instance our Seramas. They will lay well for a couple of weeks, then just stop for up to a month or two and then lay again for a couple of weeks only to stop again. This holds true for many of the more ornamental breeds of chickens. But when you raise production breeds of birds, and you know that they are supposed to keep you in eggs for most of the year, and all of a sudden stop laying, look to some form of stress as the underlying cause.

When a hen stops laying, this is an inbred instinct used as a way to protect herself and any potential offspring. Let's say that your hens are kept in an enclosed area that is fairly predator proof. Late one night, a fox comes and tries to get into the henhouse. His ruckus has sent your girls into a cackling frenzy. The fox does not manage to get in, but still the girls are spooked. It would not be unusual for some or all of your girls to stop laying for a week or two. This is because a hen, sitting on a nest to lay an egg, is an easy target. Just as a hen is about to lay her egg, she will go into an almost trance-like state. This would make her easy pickings for any predatory animal. So, to protect herself from possible threat, she will just stop laying.

Moving birds from pen to pen or to and from your farm is very stressful on a chicken. Hens have to feel comfortable with their surroundings, and any time you move them, you are placing them into an unfamiliar situation. Let's say Farmer John has a half-dozen nice laying hens that you want to buy. They are laying up a storm for him, an egg a day from each hen without fail. You purchase them, bring them home, place them in their nice clean pen and patiently wait for the eggs to fall. The next day, all the girls lay an egg just like they are supposed to. The following day, there are only two eggs and by the third day, the nests are empty. What the heck happened? You call up Farmer John and complain that he sold you defective hens.

Wait just a minute!

Those hens were completely comfortable with their surrounding over at Farmer John's. You took and moved them into unfamiliar territory. A hen can lay an egg every twenty-six to twenty-eight hours, and so your first day's eggs were already in their system when you brought them home. The second day's couple of eggs were from the least stressed hens. The third day was empty because the hen's natural instinct has kicked in and her egg producing system has shut down. Until the hens become comfortable in their new surroundings, don't expect an egg. The easier you can make the transition on them, the faster they will return to laying.

Illness also plays a big role in a hen's ability to lay. A hen in full laying production puts her body through tremendous stresses in and of itself. If your hen gets sick or is infested with parasites, her body cannot keep up with the rigors of egg laying. Her body will naturally shut down egg production in order to use the energy to fight off what is ailing her. Egg-producing hens need to be kept in top physical condition at all times.

Feed and water, or lack thereof, are huge determining factors in a hen's ability to lay. If a hen is forced to go without water for just two hours, this has the potential to shut down her egg laying. An egg is made up of nearly 90 percent water; if water is not readily available to the hen, her body cannot replenish the water that is used during the egg-making process, so it shuts down to protect the health of the hen. Likewise, feed plays a similar role. Modern feeds are specifically engineered for different stages of a chicken's life. Things as simple as changing a brand, type or quantity of feed can cause a hen to stop laying. Many people will feed their chickens all sorts of table scraps and treats. This is a perfectly acceptable way of supplementing a chicken's feed. But be aware that if you feed them too many treats and scraps, egg production is likely to drop. If you find that your egg production is slowly slipping, cut out treats for a few days and see if it doesn't rise again. I have had many people call me and tell me that their hens' production has really slowed. I ask them what they are feeding their birds and they will say that they are feeding them all sorts of scraps or bread. I simply tell them to stop the treats. And they will call back in a couple of days to tell me that egg production is right back up. Treats are fine in moderation, but if you see a drop in production, then know that you have to back off.

But by the same token, I have had people call to tell me their egg production has slipped but all they are feeding their birds is a commercially-produced layer ration. In this case I would say to step up to a better quality ration. Many times, a layer feed will just not give a hen exactly what she needs for top production. By changing feeds, many times it will be just what they need to get their bodies back into top form for egg production.

Weather and seasonal changes can also affect your hens. We have found that just before a storm, we will have great egg production. The next day will be really light. But seasonal changes have a profound effect on your hens. A hen needs a minimum of fourteen hours of light to stay in full production. As the seasons change from summer to fall and then into the throes of winter, the length of daylight becomes shorter and shorter. Hens will go into a molt and egg production will fall to near zero. This is a time of regeneration for the hens. Their bodies will use this time to recover from a long egg-laying season. They will put on weight and get all new feathers and their bodies will become ready for the next year ahead. People have found ways to manipulate this seasonal change through the use of artificial lighting, higher protein feeds, and forced molts. In the commercial egg industry, the hens used are kept in enclosed buildings, so they do not even know if it is light or dark outside and are not naturally aware of seasonal changes. They are manipulated to continue to lay for up to a year and a half without stopping. This is very hard on a hen's body and results in the death of many birds from sheer exhaustion.

All these factors are stress on a hens' bodies. If you notice a drop in egg production, look to what you can do to make their lives better and happier. Happy hens will lay happy eggs.

Genetics

Poultry genetics is a very complex subject, as it is with any species. Volumes have been written on the subject, so here we will only touch on some very basic principles.

The original chicken was the Jungle Fowl. From this breed all others were developed by way of accident or careful breeding plans. The chicken is also the closest living relative to the Tyrannosaurus rex. It is through genetics that we have been able to establish the many various breeds of poultry and to perfect the look of each breed.

Genetics come into play more so with birds used for show or production than those used for backyard pets. In the commercial layer industry, close attention is paid to those with the best laying ability. Through breeding and cross breeding, the genetic makeup of the commercial layer has been altered enough to have a bird that does not go broody and that lays a majority of its eggs within the first full year of production. In the commercial broiler industry, genetic selection has been used to produce a bird that grows to nine pounds in just eight weeks. But for most of us, the genetic makeup of our birds is more for perfecting the look of the breed.

When breeding birds, you will be best to have different bloodlines from which to work with. Line breeding is accepted to some extent, but it is not the preferred method of perpetuating a flock. It is acceptable to breed related birds if it is done as mother-son or father-daughter line breeding. If you were to breed directly related birds such as sister-brother, then you would be more apt to pull out the bad genetic traits of the birds.

The first thing that you need to establish is what trait you are trying to fix within a certain breed or bloodline. Does the comb have too many points? Then you would want to breed your birds against one that has a correct comb. Is the leg color wrong? Breed against one with the correct leg color; and so on. Many people like to experiment with color variations. This is very difficult to do and takes years of breeding before a color variation will breed true. There is a multitude of ways to go about breeding, but it all comes down to genetics.

People today are concerned with breeding heritage breeds of chickens. So many of our breeds have been so overbred to pull out one single trait or to get rid of another. This is very evident in today's commercial birds. Birds have been selected for their genetic disposition to not go broody. They have also been selected for fast growth rates. So over the years as more and more people have turned to ordering chicks through hatcheries, the ways of the heritage breeds have become a thing of the past. But through genetic selection and careful breeding programs, many of the heritage breeds of chickens are making a comeback. Hens are becoming more broody and the meat birds are growing more slowly.

Many people look to hybrid chicken breeds for their egg-laying flock. Commercial birds are also now hybrids, but on a more complex scale. A hybrid chicken will lay more eggs faster and with fewer ailments than a standard bird.

They will generally start laying at a younger age and, therefore, be more monetarily effective. But what is a hybrid chicken?

A hybrid chicken is simply a cross between two or more different breeds. There are various numbers of recognized crossbreeds that create these hybrids and usually more than one way to create the same result.

Let's look at examples of hybrids. The Black Sex-Link is created by crossing a Rhode Island Red rooster with a Barred Rock hen. The hens are all black with gold in their hackles and the roosters turn out looking like a Barred Rock with gold or orange in its hackles and saddle feathers. A Red Star is a cross between a Rhode Island Red and a White Rock.

Each of the parent breeds has its own genetic attributes, and by bringing the two together, the best of both breeds is combined to create a new breed. But with crossbred chickens of these types, they cannot be bred against each other to perpetuate the same breed. If you breed two Black Sex-Links together, the resulting offspring will revert to the Barred Rock in a sense.

Breeding for color is as confusing as it can get. There are dominant genes and recessive genes, color patterns and variations. For instance, if you breed a blue rooster against a blue hen, your resulting hatch will come out blue, black or splash.

Selecting and propagating certain genetic traits of a particular breed is much easier than color selection. There are many good books out there about poultry genetics and it would be highly advised to study one or more of these books before starting any serious breeding project. But do not be afraid of breeding your birds to obtain the best quality of structure and form according to breed specifics.

In any breeding project, do not get discouraged by the number of chicks that have to be hatched to obtain one desirable trait. If you can hatch one hundred chicks and have one that comes out with the desirable trait, then you are doing well. Perpetuating that trait is the goal. Research the history of your particular breed. Discover what original breeds were used to create the breed that you have so that you can breed back in other bloodlines with varying traits. I know this holds true with breeds such as the Delaware. They were originally known as Indian Rivers. Along the line they had New Hampshire Red bred into them. Therefore, it is possible to breed a Delaware rooster against New Hampshire Red hens and the resulting chicks come out with the Delaware color pattern. This allows you to bring in traits of another breed that was an original predecessor without affecting the outcome. This is very helpful with reestablishing body structure and leg color along with other traits such as broodiness and temperament.

Genetics and breeding can be a fun and interesting pastime. And who knows, maybe you will develop the next true breed of chicken.

SEX-LINKS AND AUTO-SEXING

Sex-linked and auto-sexed breeds of poultry are very desirable because the chicks can be distinguished at hatch as male or female. Much work has been done in the United Kingdom to establish breeds that fit this quality. For some reason, auto-sexing never really hit its stride in the United States until just recently.

There is a difference between sex-linking and auto-sexing and this can cause some confusion. The purpose of both is to be able to distinguish between male and female chicks at the time of hatch, and with both sex-link and auto-sexing, you are able to accomplish this. The difference lies in the ability to perpetuate the breed lines once established. Sex-linked chicks are a hybrid, and therefore are only first-generation crossbreeds. These resulting chicks cannot be bred back to each other to once again create a sex-link. If you bred the adult female chicks back to the original parent rooster, the resulting chicks would not be distinguishable as male and female.

With auto-sexing, when two breeds are bred together and one of them is a barred female, the resulting chicks can be distinguished at hatch as male and female, just as with sex-links. However, these chicks are purebred. So now you can take the adult chicks and breed them back to the parent stock and the resulting chicks will once again be distinguishable as male and female.

Here in the United States, we commonly have such breeds as the Black Sex-Link, Red Sex-Link, Cinnamon Queen and so on. These are sex-linked chickens. In the UK, they have such breeds as the Amrock, Barnabar, Cream Legbar and more, which are all auto-sexing breeds.

When breeding for a sex-linked chicken, two standard breeds are used. Such is the case of the Black Sex-Link, where a Rhode Island Red rooster is bred with a Barred Rock hen. The resulting chicks will be all black for the girls and black and white for the boys. As they grow, the hens will turn out mostly black with gold in their chest feathers, the roosters will look primarily like a Barred Rock, with perhaps some gold feathering in their hackle and saddle feathers. If you were to breed two Black Sex-Links back against each other, the resulting chicks would be mutts, but would look more like barred rocks.

When breeding for an auto-sexing chicken, such as the Gold Legbar, two standard breeds are also used—in this case, a Brown Leghorn rooster with a Barred Rock hen. The resulting chicks are dark crele for the males and light crele for the females. When these chicks are raised and bred against like kind, the resulting chicks will again be light and dark crele.

Auto-sexing breeds have been around for many years in the UK and elsewhere, and most have been standardized with the Poultry Club of Great Britain (which is similar to our American Poultry Association's Standard of Perfection), but auto-sexing breeds are not well known here in the United States. This is surprising due to the fact that many of the original parent stock breeds are ones that we have right here. Some of the more common sex-linked varieties of chickens are:

Black Sex-Link - Rhode Island Red rooster X Barred Rock hen

Red Sex-Link - Rhode Island Red rooster X Delaware hen

Red Star - Rhode Island Red rooster X White Leghorn hen

Cinnamon Queen - Rhode Island Red rooster X Silver Laced Wyandotte hen

Golden Comet - Rhode Island Red rooster X White Plymouth Rock hen (In all these parings, a New Hampshire rooster may be used in place of the Rhode Island Red with the same results.) Auto-sexing breeds, on the other hand, take a bit more careful breeding, and in some cases, rebreeding, before the breed holds true. But the basic parent stock is as follows:

Brussbar - Brown Sussex rooster X Barred Rock hen

Barnebar - Barnevelder rooster X Barred Rock hen

Cobar - Cochin rooster X Barred Rock hen

Gold Legbar - Brown Leghorn rooster X Barred Rock hen

Welbar - Welsummer rooster X Barred Rock hen

Wybar - Silver Laced Wyandotte rooster X Barred Rock hen

As you can see by these pairings, the common factor is the Barred Rock hen. When you put a gold gene rooster against a silver gene hen, the resulting chicks will have dark colored down for the females and yellow or light colored down for the males. But before you rush out and breed your Welsummer to your Barred Rock, be aware that none of these European breeds are yet recognized here in the United States by the American Poultry Association's Standard of Perfection. The only breed that I am aware of that is even remotely considered to be auto-sexing here in the United States is the Barred Rock—and that is because you are supposed to be able to sex a Barred Rock at hatch by the white spot on its head. A tight, well defined white spot is a female and a washed out, irregular white spot is a male.

Sex-linked and auto-sexing breeds are a great way to go if you want to guarantee that you will only have females for your upcoming layer flock. But remember with sex-linked breeds, you will need to keep both original parent stock breeds if you want future generations to perpetuate your flock. And maybe someday, if enough people work with the breeds, the auto-sexing breeds will become officially recognized here in the United States.

Sex-linked chickens are hybrids that produce a single generation chick that can be color-sexed at time of hatch. Auto-sexing chickens are true breeds that produce lines of chicks that can be color-sexed at time of hatch, generation after generation.

CROSSBREEDING

Chicken breeders sometimes find it fun to crossbreed their birds and at times come up with interesting results. This is not necessarily a good practice for a beginner, but by the same token, this is how we have established the breeds that are available throughout the world today. Just because you cross one breed with another does not mean that you end up with a new breed. What you do end up with is either a hybrid, a crossbreed or most often a mutt.

Certain crossbreeds will give you sex-linked chickens. Other crosses will give you hybrid chickens such as those used in the commercial poultry business. Or you can crossbreed to bring out certain traits in a bird. Just because you thought it would be fun to cross breed a Shamo with a Cochin does not mean that you can automatically call it a Shachin or a Como and say you have a new breed. It takes years of dedicated breeding to produce what might be considered a new breed. Once the resulting offspring can reproduce and the resulting offspring from that breeding are consistent with the parent stock, then you could present your work to the poultry community for acceptance as a possible new breed.

It is much easier to breed for certain growth or egg-laying characteristics and this is certainly an accepted practice. Take, for instance, the Cornish Cross or Rock Cross. It is a crossbreed of a standard, naturally double-breasted, male Cornish and a female of a tall, large boned strain of white Plymouth Rocks. This hybrid gives you a very fast growing chicken that is suitable for the table but is not a good egg layer. But what if you wanted a good egg layer? Maybe you would breed a Rhode Island Red with a Barred Rock and get a Black Sex-Link, which is a wonderful brown-egg-laying chicken.

Whatever your case is for wanting to crossbreed your chickens, there should be a specific goal in mind. Just because you want to see what happens by crossbreeding two different breeds is not a sufficient enough reason for doing it. All this does is create more mutt chickens in the world. The latest push in poultry breeding has been to help reestablish the heritage breeds of poultry. Some breeds, such as the Rhode Island Red, have been so overbred that it is very difficult to find good heritage stock. Just because someone says it's a Rhode Island Red does not mean that it is a true Rhode Island Red. Somewhere along the line, Johnny may have let his rooster mate with a different breed hen. The chicks hatched out and grew up looking red and so he left them in the flock. Each generation to follow had that little bit of another breed in them. After that first errant breeding, the chickens were no longer Rhode Island Reds but in fact mutts. Johnny sold the chicks as Rhode Island Reds and Timmy raised them and one day his rooster mated with a different breed hen. The chicks again hatched out and grew up looking red and so he too sold them as Rhode Island Reds. It does not take long for a breed to get so watered down that it no longer looks or acts like the original breed.

Crossbreeding poultry can bring up a lot of negative issues: one of the most dramatic is lethal genes. This is where there is a certain gene in the bird's makeup that will actually kill the bird while in the embryonic stage or sometimes shortly after hatch. Crossbreeding can also lead to many genetic defects as well as a predisposition to health problems, especially heart and neurologically related.

Crossbreeding has done great harm to the original stocks of chickens. If you are going to crossbreed your birds, then do it in a responsible manner. Do not sell the chicks or grown birds for something that they are not. If you want to be serious about it, then breed your birds to reestablish original bloodlines. Do the research necessary to understand the genetics behind what you are trying to accomplish.

Keep very accurate records of each breeding and who knows, maybe after a number of years, you may actually create a new breed that will be recognized by the poultry world.

5 POULTRY DISEASES AND AILMENTS

 ## Is There a Doctor in the House?

In the world of poultry, there are many different diseases and ailments that can pop up unexpectedly and rapidly. Most have a cure but some only have a treatment. Poultry have long been considered a disposable commodity with little knowledge set forth to their immediate care. Granny had her homespun cure-alls for her little barnyard flock; some worked and many were just old wives' tales. With the advent of the commercial poultry industry, it was not economically feasible to waste a bunch of time, energy and money on curing a sick bird, so they were just culled out of the flock.

Chickens and the backyard flock have taken a new turn. Now, many birds are being kept as pets, and so there is a new focus on the care and well being of individual birds. But accurate diagnosis and treatment plans are still very archaic at best. There are very few poultry-specific medications available. Home diagnosis of diseases is not always possible and finding a local vet who is highly trained in poultry medicine is nearly impossible as well as very expensive. So we, as backyard poultry enthusiasts, are left to our own devices to determine what ailment our birds might have and what course of action best suits the problem. Many times we are left with heartbreaking results. But we, as a poultry community, cannot give up, and we must be willing to share information with one another.

There are many problems that arise from diagnosing and determining treatment for an ailing chicken. Many disease symptoms mimic those of others, and so, without laboratory services, diagnosis becomes a guess at best. Another problem is that poultry can harbor a disease for great lengths of time before it ever shows symptoms, and by the time you recognize that there is a problem, it can very easily be too

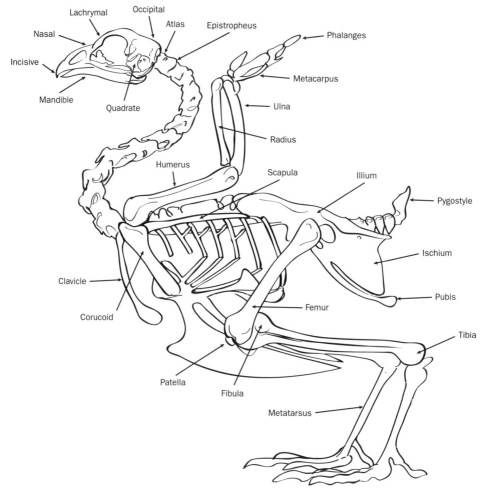

Poultry skeleton

late to treat it. Some diseases have such a rapid onset that if you are not fully prepared, you can lose a number of birds before you even have time to treat the problem.

To further complicate matters is the mode of transmission. Some diseases are passed only by direct contact. Some are passed by indirect contact, meaning that if a sick bird drinks from a waterer and then another bird comes and drinks from the same water, then it can catch whatever the ailment may be. And yet other diseases can be passed on the wind. You could have a farmer miles away and his birds could

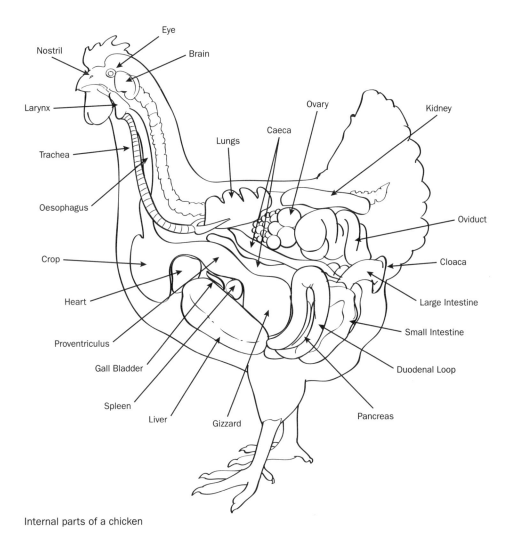

Internal parts of a chicken

develop a disease and it can be carried on the wind to your flock. You could also go visit another poultry grower and handle his birds and come home and handle your birds and inadvertently spread disease. If you are going to have chickens, it doesn't matter how careful you are, they are always susceptible to disease. You should never let your guard down. It is always important to do what you can to eliminate as many possible modes of transmission as possible, but just know that you have no way of always keeping your birds disease-free. There is always that chance. The secret is to start with healthy birds, use disinfectants and sanitizers regularly and watch your birds for the first signs of problems.

Many of the medications used today to treat poultry ailments are actually designed for larger livestock animals. This creates a series of problems in that, even if you can diagnose the disease, it becomes a matter of what medication to use and in what dosage. Many times you find yourself having to use a medication in a way that is not listed on the label. This is what is known as "off-label use." In the world of poultry, this off-label use creates a new problem in the fact that chickens are generally kept for the purpose of eggs and meat. If you use a medication in an off-label way, then there has not been research done to determine the withdrawal time for the safe consumption of the eggs or meat of that bird. Because the egg is formed inside the hen's body over a period of twenty-six to twenty-eight hours, this gives time for medications to be secreted into different layers of the egg. If you were to use a medication on your bird that you in fact are allergic to and then consumed the egg, there is a possibility that there would be a sufficient amount of that medication in the egg to create an allergic reaction.

But this off-label use of medications in the treatment of poultry has also led to a secretive world among poultry growers where they are afraid to share information because of our fine government policies. The agricultural folks watch the forum sites to see what people are doing in the ways of off-label treatments of animals. From this information they can and do locate the people and inspect their farms for illegal use of medications because poultry are a consumable product. In a way I understand their point, but there is also the point of crossing the line. The government does not do enough to work with, and share with, the backyard farmer on viable treatments for poultry ailments. So the whole process is slowed by lack of research as well as an unwillingness to share information.

For most medications there is an unwritten standard within the chicken community that a withdrawal time of ten days is generally sufficient, which is to say that you would not eat the eggs or meat of the chicken during, and for ten days after, the treatment. But you will also hear from different people that they do not heed the recommended withdrawal times simply because they feel what little medication may be transferred to an egg or retained in the meat will only benefit them in some way.

Take worming, for instance. It is a well-known medical fact that humans can also have worms. People who worm their chickens feel that if they happen to ingest a bit of wormer, then it is only benefiting them by killing some of their own worms. Many people use a topical wormer that is absorbed into the skin of the chicken. This is very difficult to apply without getting any on yourself. With all the chickens that I have wormed, I feel I am worm free for the rest of my life.

There are many misconceptions about poultry diseases. Most are not transmittable to humans. There are the rare few that in certain cases could be passed to humans, but you are not going to get chicken pox from a chicken. Yes, a chicken may develop fowl pox and yes, it is of the same type of virus, but it is not the same strain of virus. But it is possible for you to get secondary infections from poultry such as Salmonella, E. coli or Staphylococcus. Salmonella and E. coli are easily transmitted to humans through the ingestion of fecal matter. Yes, it sounds gross, but chickens poop and they walk in it and if you handle your chickens then you are going to get fecal matter on your hands. Therefore, it is very important to always wash and sanitize your hands after handling your birds and before eating.

Staphylococcus, or Staph, is present everywhere in our everyday lives. We are constantly exposed to it even in what is supposed to be the fairly sterile atmosphere of a hospital. Chickens are very susceptible to Staph infections and so it is very important that if you are doing any type of surgical procedure on your birds that you wear rubber gloves of some sort. It is also very important, once again, to always wash and disinfect your hands after handling your birds. But also just as important is that a chicken can carry the Staph germs on its feet and body, so if it happens to scratch you, it is important that you properly treat the wound immediately.

Do not get discouraged if one of your birds becomes ill. Try to catch the symptoms early so that you can provide better treatment faster. Chickens are very resilient creatures and usually pull through fine. Learn from everything that may come up and be as prepared as possible for a wide variety of possible problems. Realize that many times you will be left to your own knowledge and expertise to diagnose and treat your poultry. Do your best at finding a remedy that works and remember it for the next time.

But with all that said, chickens are relatively clean animals. Their cleanliness is a byproduct of what type of conditions you keep your birds in. If they are allowed to be kept in filthy, poop-covered pens, then you are going to have many more disease problems than if they are kept in relatively clean conditions. It really is not hard to keep your birds in a clean, safe, healthy environment, and you and your birds will greatly benefit from it.

Although there are many different diseases that poultry are susceptible to, in the following section I will try to cover some of the more common diseases and ailments that might someday afflict your birds. Be aware that some of the pictures may be quite graphic in nature and not suitable for all ages without proper explanation. But also be aware that we are dealing with a living creature and living creatures can and will get sick and sometimes die. It is a reality and one that must be faced. It is my hope that through this next section that you will be better able to understand the possibilities that may arise and handle them in a way that is fitting to the situation.

Three-Legged Chickens

A man was driving along a freeway when he noticed a chicken running alongside his car. He was amazed to see the chicken keeping up with him, as he was doing 50 miles per hour. He accelerated to 60, and the chicken stayed right next to him. He sped up to 75 miles per hour, and the chicken passed him.

The man noticed that the chicken had three legs. So he followed the chicken down a road and ended up at a farm. He got out of his car and saw that all the chickens had three legs.

Man: What's up with these chickens?

Farmer: Well, everybody likes chicken legs, so I bred a three-legged bird. I'm going to be a millionaire.

Man: How do they taste?

Farmer: Don't know, haven't caught one yet.

Parasites

Chickens peck the ground and just about anything else. Parasites live in the dirt and just about everywhere else. Therefore, the chances of you having to worm your chickens at some point are pretty great. How to go about getting rid of them can stir up some great debate.

WORMS

You will need to worm your birds if they are over six months old and have been on dirt, if your bird has an unknown worming history, if your bird has not been wormed in nine months, if you see any worms in the fecal matter, if you are experiencing low feed-to-weight conversion or if your birds are losing weight unexpectedly.

What works for us is this; any birds that we raise from chicks first get wormed at six months of age and then every six months after. Any birds that are brought in are wormed right away and then every six months after. We generally follow a twice a year program as we do not seem to have the problems that others do with worm infestation. Many people will only worm their birds once a year. You have to establish what works best for you and your area.

Our first worming is always done with Wazine 17 at a rate of two tablespoons per gallon of water. This is given to the birds as their only source of water for twenty-four hours. After the twenty-four hours, they go back to drinking plain water. This is followed up two weeks later with Ivermectin Pour-On cattle wormer placed on the skin at the back of their neck at a rate of eight drops per full-size bird and accordingly less per smaller size birds; our small bantams receive five drops.

So, you ask, why the double worming? Wazine 17 is a milder wormer that will only kill round worms and nodular worms. Ivermectin Pour-On is a broad spectrum anti-parasitic that will kill everything. If your bird happens to be heavily infested with worms and you only used the Wazine, you would not kill them all. If you only used the Ivermectin, then when all the worms die, it can flood the bird's system with too much foreign protein and you risk killing your bird. After the initial worming with Wazine, then you can use just the Ivermectin at each worming after that. On rare occasions we find that we have to worm our birds at closer intervals. With each worming we use a ten-day withdrawal time of the eggs. Many people do not do this with the thinking that if they eat the eggs, then they too will be rid of any parasites.

There are many different anti-parasitic medications out there and people all have their different preferences. There is fenbendazole, sold under the trade name SafeGuard or Panacur. This is a paste that you give the bird in a piece the size of a BB to eat. There is also levamisole and Tramisol. There is the more natural route of giving your birds apple cider vinegar, red pepper flakes or diatomaceous earth. It is all in the way that you wish to raise your birds. We have found that the pepper flakes and the diatomaceous earth are fine for a maintenance program, but neither, in our opinion, does enough to clear the system of all parasites.

Whichever route you choose to follow is up to you, but birds are very susceptible to worms and parasites, so it is in your best interest, as well as your birds', to get them on a worming maintenance program and stick with it. You and your birds will be happier because you did.

MITES AND LICE

Mites and lice are two of the most common parasites on poultry. For many people, just thinking about them makes their skin itch. Well, if just thinking about them makes your skin itch, think about what the actual thing is doing for your birds. They are not hard to keep under control if detected early, but severe infestations can become frustrating to treat and eradicate. Through simple, periodic examinations of your birds and easy treatment procedures, you and your birds can remain healthy and happy.

Poultry lice are tiny. They are about a tenth of an inch long and kind of yellowish in color. They spend their entire life on the birds and can be passed from one bird to another by close contact. Poultry lice are host specific, meaning that they can live on only one species of animal. Poultry lice cannot survive on humans, nor will they infest your dog or cat.

They generally feed on dead skin cells, scabs and parts of the feather. They will occasionally feed on blood from open sores. The poultry louse will lay between fifty and three hundred eggs and attach them to the base of the feather shaft.

Examination for lice is quite simple. We find it is easiest at night when the birds are roosting to do our examinations for lice and mites. Pick your bird off the roost and cradle it in your arm much like you would be holding a football, with its head under your arm. With a helper aiming the flashlight, pull back the feathers around the vent area and look for lice crawling around the skin. They move pretty fast so you have to look closely. Pull back different areas of feathers to continue your examination. Now check around the thighs and breast. Also look at the base of the feather shafts for clusters of eggs. Your greatest infestations of lice are going to happen in the fall and winter.

Mites, on the other hand, are even smaller yet. They appear red or black and look like tiny moving specks. There are two main types of mites, the northern fowl mite (or in tropical regions would be called the tropical fowl mite) and the red poultry mite (or red roost mite). Mites are different from lice in the fact that they suck blood from the host animal. The northern fowl mite will spend most, if not all, of its life on the host animal. The red mite is a nocturnal feeder and will infest the bird at night to feed and then leave the host animal to hide in cracks and crevices of the poultry house during the day. Because mites are most active at night, nighttime examination is best to determine the level of infestation of both mites and lice. Like lice, mites also prefer the vent region amongst the soft feathers. Also check the areas under the wings in the short, soft body feathers.

If mites and lice are discovered on your birds, treatment is fairly easy depending on the level of infestation. Your birds will take a dust bath usually once a day. This is their natural way of ridding their bodies of parasites along with removing extra oils from their feathers. Use their dust bathing to your advantage by adding a bit of Sevin dust or diatomaceous earth to their dust bathing area. As the birds works the dust through their feathers, they will also be helping to kill the parasites. For more severe infestations, it may be necessary to use a pesticide directly on their body. Products such as permethin, Pyrethrum, carbaryl or Sevin dust work well. If treating for red mites, be sure to spray down the entire housing area, paying close attention to cracks and crevices where the mites could hide. Depending on the level of infestation, retreatment may be necessary to completely break the cycle.

In the case of heavy infestation, it may be necessary to contact your local avian veterinarian about treatment with products such as Ivermectin. Ivermectin is a wonderful product for the treatment of parasites, both external and internal, but should only be used under the supervision of a veterinarian because withdrawal times have not been established. Ivermectin comes in various forms, but the most commonly used for parasite treatment is Ivermectin Pour-On for cattle. This is a blue-colored, alcohol-based cattle wormer. On poultry, it is given at a dose rate of eight drops on the skin at the back of the neck for large fowl breeds. The bantam rate is generally five drops. This is absorbed into the skin and distributed throughout the body through the blood system. This will kill both external and internal parasites.

Another common, and equally effective, veterinarian supervised treatment for mites and lice is the use of Frontline or FiproGuard for dogs. This is an oil-based anti-parasitic that only treats external parasites. It is generally dosed at a rate of three drops at the vent area and two drops on each inner thigh. It is spread about the skin through normal grooming and works rapidly to control infestations and has great residual effect for up to thirty days, although treatment is usually repeated after fourteen days. For heavy infestations, both Ivermectin and FiproGuard are great treatments, but should be supervised by a trained veterinarian. Additional treatment of the ground and facility will still need to be performed using a pesticidal spray to kill any lingering parasites.

SCALY LEG MITES

Scaly leg mites are a common problem with chickens and we are all faced with them at some time or another. If you are out in your chicken pen and notice raised scales on your chickens' feet or legs, then you can pretty much count on them being affected by scaly leg mites. These mites can cause your chicken to have soreness, irritation, pain and possibly even lameness. This pain and irritation can lead to other injuries and severe infections, and in severe cases could lead to death.

There are numerous treatments for scaly leg mites, ranging from the old harsh remedy of dipping the chickens' legs and feet in warm kerosene to the more modern ways of anti-parasitic medications.

Petroleum Jelly (Vaseline)

This is the standby for most people. It will suffocate the mites but is not very effective on feather-legged birds or birds that have developed thick scales. You can mix in a bit of Sevin dust with the petroleum jelly if you choose. Smear on the petroleum jelly and push it up under the scales. This will have to be repeated every day for at least two weeks to be sure to get the adults as well as the new hatchlings.

Campho-Phenique

This is a great product and is available at most pharmacies and grocery stores. Holding your bird in a slightly upside-down position, dribble the Campho-Phenique on the scaly areas of the legs and feet, allowing it to run up under the scales. Massage it over the entire leg and foot. This will drown and kill the mites. Campho-Phenique will also help prevent and cure infection as well as provide the bird with a bit of safe painkiller to help ease the irritation. Apply this treatment twice a week for three weeks to be sure to get the adults and also the newly hatching mites.

Ivermectin Pour-On for Cattle

If you already have your birds on a semi-annual worming program, chances are you are using Ivermectin as your wormer of choice. Chickens that are on a regular worming schedule that includes Ivermectin will generally not get scaly leg mites. Ivermectin Pour-On is an anti-parasitic that is absorbed through the skin and is effective on almost all parasites including the scaly leg mites. If you do not already use Ivermectin or you notice an infestation even though you are using Ivermectin, you can place three drops of Ivermectin at the top of each leg right below where the feathers start. Ivermectin will work quickly to kill the mites and will have enough residual effect to kill any hatchlings, though it is advisable to treat the legs again after two weeks just to be sure that all the hatchlings are killed.

Did you know?

Chickens are wonderful animals to travel with. Once in motion, they will lie down and be fairly quiet until the next stop. Place a piece of fruit in the cage so that they may eat and get hydration along the trip.

Frontline or FiproGuard

Used only on the advice of your poultry veterinarian. Frontline and FiproGuard are the same type of product with the exact same chemical makeup, but as of this writing, FiproGuard is much cheaper yet just as effective. FiproGuard is a topical anti-parasitic that is placed on the skin of a large fowl at a rate of three drops on each inner thigh and two drops just below the vent. On bantam breeds, the rate of application would be two drops on each inner thigh and one drop below the vent. Generally, only one application is needed to kill the adult mites as well as any eggs that hatch. Just as a side note, FiproGuard also works very well on northern fowl mites.

With any of these treatments, no egg withdrawal times have been established, though most recommend a ten-day withdrawal from eating eggs. It would be advised to check with your poultry veterinarian as to egg and meat withdrawal times.

As with any treatment, you will need to be aware of the living conditions of the bird and correct any problems by changing the litter, using a good insecticide to kill any residual bugs in the living quarters and maintaining a watchful eye for any re-infestation. Scaly leg mites can be very painful and annoying to the chicken, but with proper treatment can be cured and the bird can return to a normal way of life.

STICK-TIGHT FLEAS

Stick-tight fleas are the peskiest little parasites I have ever had a chance to witness first hand. Stick-tight fleas are naturally occurring parasites that inhabit the warm, humid southern states, primarily Florida, Texas and California. They are found on mice, rats, dogs, cats and many other fur- and feather-bearing animals.

It is the females that attach themselves to the host animal primarily around the eyes and ears much like a tick does. If left alone, they will strip all the feathers off the heads of your birds, and it usually takes two good molts for the feathers to grow back in. Stick-tight fleas are dark in color and are really tiny. They will cluster around the head and look like dirt at first. If you look really close you can actually see them moving. You can try to scrape them off with your fingernail but they are stuck so tight that that usually does no good, hence the name.

The females will lay their eggs, which fall to the ground. The males remain in the dirt or sand and fertilize the eggs as they fall. The eggs pupate and the larvae will cocoon in dirt and lay dormant until disturbed and can hatch out at anytime up to twenty-one days. The females will then jump to the host animal to continue the infestation. The birds' scratching around will stir up the fleas, as will you just walking into their pen.

Stick-tight fleas are a seasonal parasite just as regular fleas tend to be. They are not known to transmit any diseases but they can cause severe secondary infections at the attachment site. Weather patterns of consistent rain and high humidity followed by drying trends tend to heighten the presence of the parasite. Little is known on how to eradicate this pest other than a constant regimen of parasite control. They seem to be quite territorial. This is to say that once they infest an area they do not migrate far unless spread by some mechanical means such as on shoes or clothing or by moving birds from one area to another.

If caught in the early stages, they are not too hard to eradicate. But left unchecked, they can be horribly frustrating to get rid of. If you find yourself in a fight against these pests, there are a few things that will help. It seems that these fleas cannot jump higher than three feet. If you have only a few birds and can put them in elevated pens, this will greatly help. Elevate them above three feet and it will help

Stick-tight fleas are a seasonal parasite just as regular fleas tend to be. They are not known to transmit any diseases but they can cause severe secondary infections at the attachment site. They will cluster around the head and look like dirt at first. If you look really close you can actually see them moving.

keep them from being further infested. If you have a lot of birds, this could be nearly impossible to do. You can also cover the infested area on the host animal with petroleum jelly (Vaseline) to smother the fleas. Use 5 percent or 7 percent Sevin dust on the ground and surrounding areas. The use of permethrin in a spray is also beneficial to help kill the ground-based parasites. The use of typical anti-parasitic drugs, such as Ivermectin, seems to offer only temporary relief.

The best and most effective treatment that we have found is to clean out all nesting materials and spray the nest boxes with a permethrin-based product. Spray down the pen area up to a height of three feet, paying close attention to cracks or areas where the fleas could hide. Talk with your avian vet about the use of Frontline or FiproGuard on your birds. This has had great success as on off-label remedy with quick results. FiproGuard is used at a rate of two drops per bird placed at the back base of the comb. This is repeated after two days. Continue to check the birds carefully for continued infestation. This can be repeated after one week if necessary.

Frontline or FiproGuard have not been approved yet for open use on poultry, so egg and meat withdrawal times would have to be established by your vet.

While dealing with an infestation of stick-tight fleas, be sure to check your body carefully when finished in the pens, as they will also bite humans, though it is not common.

If not controlled, stick-tight fleas will cause your birds to become anemic. They will slow or stop laying and your birds will slow their feed and water consumption and eventually die from malnutrition.

Stick-tight fleas are a formidable pest and it takes great diligence to rid your birds and property of them.

Did you know?

Chickens are very clean animals. They constantly groom themselves, take baths regularly, and with a bit of light maintenance from us, can be as clean or cleaner than most other household pets.

Ailments

There are a multitude of ailments that a chicken may develop over its lifetime. Most of these are related to vitamin deficiencies or genetic factors but can also be caused by problems during incubation. The most common and simplest ailments tend to affect chicks and young birds the most. A chicken's body grows very rapidly within the first six months, so the opportunity for nutritional or genetic issues to show themselves is much greater. Once a bird reaches the adult stage, there is not such a prevalence of ailments that arise. The following is a short list of the most likely ailments you may witness in growing birds.

CURLED TOES

Once in a while, when a chick hatches, it will have curled toes on one foot or both. Curled toes are generally caused by the humidity being too low during incubation. It can also be caused by a genetic factor, but that is rare.

This is usually a very easy deformity to correct if caught early enough. A newborn chick's bones are still very soft just after hatching but begin to harden within the first forty-eight hours and are fully hardened within five days. There are many different remedies to this and all are pretty close in style, but we use Band-Aid adhesive bandages the most often with great results.

For best results, this needs to be done within the first twenty-four hours. After the chick has dried and is all fluffy, get yourself a couple of larger adhesive bandages. Depending on how big the chick is will determine how big of a bandage you will need. We generally use the ones that are one inch wide. Cut the pad part out of the bandage so you are left with the two sticky ends. Holding the chick in one hand, place its foot onto one of the sticky ends of the bandage. Straighten out its toes so they are stuck in the right position.

Now take the other bandage sticky end and place it over the top of the foot and press the two together, especially in between the toes.

Take a small pair of scissors and trim off any extra tape. Your chick will now have duck feet. Place the chick back into the brooder. It will have trouble walking at first, but it will soon figure it out and be up running around with all the others. Leave this on for twenty-four hours. After twenty-four hours, gently remove the tape and see how the toes look. If they are still curled at all, repeat the procedure for another twenty-four hours. Usually their toes will be cured with the first application. This procedure is much easier to accomplish with the aid of another person, but it is certainly workable by yourself.

Be extra careful when removing the adhesive bandages as you can peel the skin and remove toenails if you are not gentle. We find it is easiest, when removing the bandages, to cut between and as close to the toes as possible. Then wet the bandage before gently peeling the pieces off.

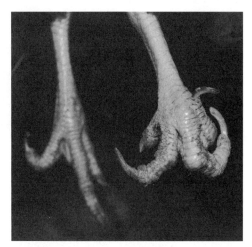

Once in a while a chick will hatch with curled toes. Curled toes are generally caused by the humidity being too low during incubation or by a rare genetic factor.

If you have an older chick that develops curled toes, this is caused by a vitamin deficiency, specifically riboflavin, and can usually be corrected by giving added vitamins and minerals in the water to include riboflavin. In addition to the curled toes, chicks affected with this vitamin deficiency will walk on their hocks and appear unable to stand on their feet. Riboflavin deficiencies can be passed from the hen through the egg, which affects the growth of the embryo.

In any case, it is imperative that quick action is taken to correct the problem, or permanent and debilitating damage to the toes will occur.

Within 24 hours after the chick is fluffy and dry, a remedy using adhesive bandages can be used for curled toes.

PASTY BUTT

It is all too common for your chicks to get pasty butt, where their poop dries on their butt. This can cause serious problems if it should block their vent. The poop can back up into their system and cause infections and even death. It is very important to keep a watchful eye on all your chicks and take care of the problem immediately.

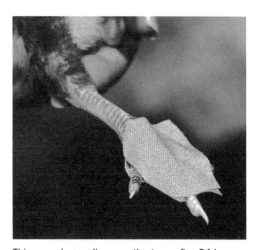

This remedy usually cures the toes after 24 hours. Quick action must be taken to correct the problem to avoid permanent and debilitating damage to the chicken's toes.

The easiest way to handle this is to hold the chick firmly in one hand and use your thumbnail against your first finger to pick it off. I know it sounds gross but you can wash your hands afterwards. Be careful when doing this, as the vent area is very tender and sensitive in chickens. Slowly pick away the dried poop until most of it is gone. You can then take a warm, wet washcloth, just above body temperature,

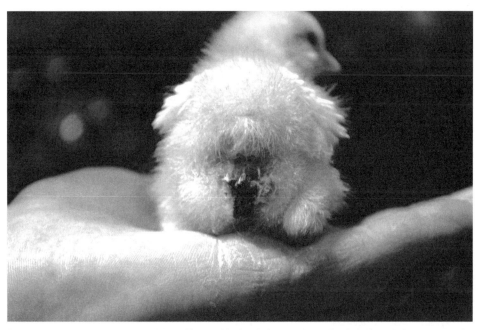

Pasty butt can cause serious problems if poop blocks their vent. It can back up into their system and cause infections and even death.

and gently wash the area. Then dry its little butt and put it back into the brooder to completely dry and warm back up. Be very careful that you don't pull out the downy fluff when you are doing this. You could make the chick start to bleed and then you would have to put a bit of blood stop or other coagulant on the area to stop the bleeding. If this should occur, keep the chick separated from the others until it heals or the other chicks could start picking at it, which could potentially kill it. Once a chick learns to squat better, then the instances of pasty butt will go down.

Pasty butt also happens with older chickens, especially hens. This is usually caused by too much butt fluff around the vent, that is those fluffy feathers around its butt that make it look so big from the back. This is handled pretty much the same ways as in chicks. We firmly hold the offending chicken in one hand with its head tucked under our arm. Then we take a pair of sharp scissors and slowly cut away some of the crusty fluff from around the vent area. This is just to trim back some of the feathers so there is less chance of a repeat offense. Be very careful that you don't cut too much of the feather away or it will cause bleeding. Don't cut more than about half of the feather shaft. Once most of the poopy fluff is cut away, then you can gently wash the vent area with warm water. Dry the area and do any final trimming of feathers if necessary.

SLIPPED ACHILLES TENDON

On occasion, a chick will be born with or will develop a slipped Achilles tendon. The Achilles tendon runs down the back of the chick's leg and, at the hock joint, rides in a groove. Because a chick's bones are rapidly growing and forming when it is born, it is possible that the groove has not yet fully formed enough to properly hold the tendon in place. If the tendon slips out of this groove, the joint will look swollen with the back of the hock joint appearing flat. The chick will be unable to straighten that leg and, therefore, will be unable to walk. This is generally an easy problem to fix.

While holding the chick, gently pull the upper part of the chick's leg a bit behind the normal position, and then carefully straighten the whole leg as though the chick was stretching its leg backwards. The tendon should pop back into place. If it doesn't pop into place on its own, slight pressure with a finger will usually do the trick. Allow the chick to bend its leg back to a normal position and see if the tendon stays in place. Depending on how well formed the joint is, you may have to repeat this procedure to get the tendon to stay in place.

Once the tendon remains in place, a splint or cast can be made from a bendable drinking straw to help protect the joint and tendon while it heals.

To make a splint, cut the bendable part from a drinking straw slightly above and below the corrugations. Now split the bendable part of the straw lengthwise. Cut the length of the splint if necessary to match the length of the leg. Place the bendable splint around the leg with the slit to the front and place tape around the splint to hold it together and in place. The tendon will need to be checked periodically to make sure it is in proper alignment. Leave the splint on for four to five days until the joint and tendon have healed.

SPRADDLE LEG

Spraddle leg or splayed leg is a condition where when the chick is born its legs are not in the right position and it looks bow-legged or one leg or both are splayed out to the sides and the chick has trouble walking. Newborn chicks can also develop spraddle leg if the floor of your brooder is slippery. We find it best to use a couple of layers of paper towels on the floor of the brooder for the first four to five days to help keep the chicks from sliding and causing damage to their sensitive bones.

Spraddle leg is generally an easy condition to correct if caught within the first twenty-four hours. A newborn chick's bones are still very soft just after hatching, but will begin to harden within the first forty-eight hours. Its muscle development is also rapidly happening at this time. If left longer than twenty-four hours, there is a good chance that you will not be able to remedy the problem and your chick will be permanently crippled.

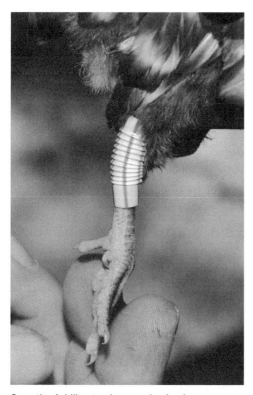

Once the Achilles tendon remains in place, a splint or cast can be made from a bendable drinking straw to help protect the joint and tendon while it heals.

There are many different remedies for this but we have found that using pipe cleaners is the simplest way to correct the problem. Cut a length of pipe cleaner about three inches long. Wrap one end of the pipe cleaner around one leg two or three times. Spacing the legs about one inch apart, wrap the other end of the pipe cleaner around the other leg two or three times. Cut off any excess length.

Leave this hobble on for twenty-four hours. The chick will have difficulty walking at first but will soon figure it out. You may find that you have to separate the chick from the others while it is hobbled because the others will pick at the hobble out of curiosity. After twenty-four hours, remove the pipe cleaner and see if the legs remain in the proper position. If not, then place the hobble on for another twenty-four hours until the legs are held in the correct position.

An alternate way to correct this problem is to use small Band-Aid adhesive bandages. For large fowl chicks we find that the small half-inch-wide bandages work the best. We use three adhesive bandages for this procedure. Cut the pad out of one bandage so you are left with the two tape strips. This procedure is most easily handled with two people, one to hold the chick and one to apply the bandages. While one person is holding the chick, take an adhesive bandage and wrap it around the leg so that the pad portion is around the leg and stick the two tab ends together facing towards the middle. Repeat the procedure on the other leg. Now take one of the tape sections that you cut off the third bandage and, while holding the tab ends together from the two bandages that are wrapped around the legs, make sure the legs are in a normal position and tape the bandages together. Use the other tape end to tape the other side just as reinforcement. Place the chick back into the brooder. Leave this hobble on for twenty-four hours. The chick will have trouble walking at first but they usually figure out how to walk after a while. Make sure that the other chicks do not trample it. After twenty-four hours, remove the tape strips, leaving the other bandages in place, and see how the chick walks. If the legs are still splayed, then reapply new tape strips to hold the legs in position for another twenty-four hours. This condition is usually cured within the first twenty-four hours.

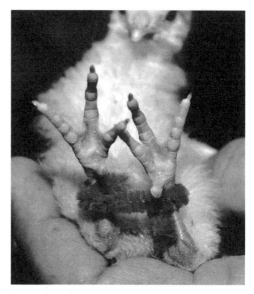

To correct spraddle leg, cut a length of pipe cleaner and wrap one end around one leg two or three times. Spacing the legs one inch apart, wrap the other end of the pipe cleaner around the other leg. Leave this hobble on for 24 hours.

VITAMIN AND MINERAL DEFICIENCIES

Today's commercial feeds are designed to provide chickens with all the necessary vitamins and minerals that your birds need at different stages of their lives. But there is the occasion where one or more birds within the flock may not be able to process enough of one or more certain vitamins or minerals from their daily feed. This can result in different ailments or deformities manifesting themselves within your flock.

Some deficiency symptoms also mimic symptoms of other diseases, so it becomes difficult to determine an exact cause.

Birds who are free-ranged, given large amounts of household scraps, or fed custom diets are much more susceptible to vitamin and mineral deficiencies than those kept on a strict diet of commercially produced feeds. As with all feeding programs, the levels of absorbable vitamins and minerals must be closely monitored. The advantage of feeding commercial feeds is that extensive research has gone into the production of the feeds to provide the correct amounts necessary for optimum growth and production. But this is not to say that all feeds are created equal. There are plenty of low-quality feeds being produced and sold through feed stores, and your birds' health and vitality will be directly affected.

When starting out with a new flock, don't be afraid to change feeds now and again until you find a feed that suits your birds well for the environmental conditions that they are living in. Young chicks require a slightly higher protein level with lower levels of calcium, whereas a meat bird will require much higher levels of protein and laying hens will require sufficient levels of protein with added calcium for good egg production. But in this mix, a deficiency of one or more vitamins and minerals can occur. This is especially true if you raise birds of mixed age and body type together.

There is a new wave of thinking amongst poultry growers, especially the folks that have a small backyard flock, to raise their birds on the most natural diet possible. I certainly understand the thinking of wanting to get away from the commercially produced feeds, where most of the ingredients cannot even be pronounced, much like our own human foods. This is a noble approach to raising healthy birds, for you and for them, but there is also the problem of supplying your birds with the correct nutritional needs. Today's birds are a far cry from the original Jungle Fowl from which all breeds originated. The breeds today have been so overbred to bring out specific traits that most are no longer good foragers in a free-range environment. The push with selective breeding to create fast-growing meat birds or the super layers have pretty much left these birds to the mercy of the commercial feed producing companies. But there is also a new way of thinking that has breeders now breeding the other way, to get certain breeds back to what they were many years ago. Meat birds are being allowed to grow a bit slower and layers are not expected to lay three hundred eggs a year.

There are now birds being bred, known as Freedom Rangers, that are great foragers and so require less commercial feeds yet still produce adequate supplies of eggs and nice Sunday dinners. Breeders are working hard to bring back many of the heritage breeds of chickens as well, the ones like Grandma raised on her farm. People are turning away from the store-bought fryers that reach eight pounds in eight weeks in favor of homegrown birds that reach a mature weight in fourteen to sixteen weeks or longer. People are also turning away from store-bought eggs that are watery in consistency, spread out flat in the pan and are laid by hens that are kept under deplorable conditions in battery cages. People are coming back to the old ways of thinking, mainly for the health of their chickens and to know what is going into the foods that go onto their table.

But by doing this, people are also discovering that the healthier way of living is also a bit more complex than they first imagined. To keep your birds healthy and happy, you must be able to provide them with all that they need, especially in the nutritional aspect. There are thirteen vitamins and twelve minerals that directly affect the performance of your chickens. A deficiency in any one of these can cause significant problems, but most are readily remedied.

FLIP-OVER SYNDROME

Not much is known about this strange phenomenon in the backyard poultry world and a search on the Internet will yield little or no information. Yet this does happen, and few people ever realize that something is different.

When a chicken dies from a non-predatory attack, it will generally die on its side or on it's belly. When a chicken dies from flip-over syndrome, it dies flat on its back. Death is nearly instantaneous with no warning signs.

I have witnessed this phenomenon firsthand and it is quite remarkable. A hen or rooster who appears perfectly healthy, eating and scratching, all of a sudden stands upright, flaps its wings and tips over backwards dead. From start to finish it is generally less than sixty seconds. The post-mortem signs are very atypical in that the bird is flat on its back with wings out to each side. The head is laid straight back and the feet are up in the air. It is almost like a stereotypical comic death pose. In dealing with a couple of large broiler poultry houses, the subject has come up. Through this I have learned that this condition is fairly common among fast-growing birds. It is directly related to heart conditions. In fast-growing birds, where the bird is reaching a market weight of eight pounds in just eight weeks, the heart is put under great stress. The bird will easily die of a heart attack. This appears to be where this syndrome comes from. They refer to it as "flippers."

In a small backyard flock of laying hens, this is much more uncommon but it does happen. It could be understood in the large broiler houses where the birds' bodies are put under so much stress, but in a backyard flock their bodies are not under those types of stressful conditions. It is thought that there could be a genetic factor at play. Just as if your parents have heart-related problems, so could you be more susceptible to heart problems.

It appears that this is caused by the bird having a heart attack, but little post-mortem research has been done in this area. What they have found is that this is a disease that predominantly affects well-developed male birds. The large poultry houses just dispose of the birds and in a backyard flock most people just figure a bird died so they bury it. This, along with the fact that poultry houses are not going to take the time to do necropsies and most private poultry people cannot afford a necropsy, there is little known about the direct cause of flip-over syndrome.

There is no known preventative for this. No warning signs or symptoms. It is just one of those things that happens, unfortunately. So, if you do have a bird that all of a sudden drops dead and is laying flat on its back, don't feel that you have done something wrong. It is not contagious and will not affect any other members of your flock. If you have other birds from the same clutch of chicks, it is possible that more than one could have the same result. It would not be advisable to heavily breed birds that are predisposed to this condition as their offspring could also be at risk. But the risk and rate of occurrence is extremely low. Out of more than 3,000 birds that have come and gone through our farm, we only know of three that have died this way. And in the large poultry houses where they are dealing with hundreds of thousands of birds in extremely stressful conditions, they may lose a hundred or so birds to this condition during each eight-week contract. It is not something to worry about, but just something to be aware of.

In our small backyard flocks, every death is a major ordeal and we rush to make sure that the others are not going to be caused the same fate by something that we have done or not done. This is one of those instances where there is nothing you could have done either way.

WATER BELLY

Water belly refers to the fluid accumulated in the abdominal cavity as a consequence of heart failure. The disease is more scientifically known as pulmonary hypertension syndrome and this disease may or may not actually end up as what is known as ascites.

This condition is most often found in rapidly grown meat birds, yet it can affect any bird. It is a disease that is brought on by the failure of the right ventricle of the heart. In humans, this is more commonly known as congestive heart failure, where fluid builds up around the heart and lungs.

A similar, though more complex, situation will also occur in the chicken's circulatory system. The blood pressure will back up from the lungs, through the heart, back to the liver and abdominal viscera. The heart will enlarge due to both pressure and, with time, hypertrophy of the muscles due to the hard pumping activity (exactly the way in which skeletal muscles will enlarge in weight lifters as they "pump iron"). Further upstream, the blood vessels will also enlarge, causing the liver to swell and blood vessels on the intestines to become prominent. Because blood vessels are also quite leaky, fluid will leak out and that is the source of all the extra "water" seen in the abdomen of affected birds.

In most backyard poultry, this condition can be maintained through proper diet. Individual birds that show swelling of the abdomen can be drained to relieve pressure and distress. As the abdomen swells it makes it more difficult for the bird to walk. It also puts pressure on the internal organs. Draining the abdomen of a bird is actually a simple procedure. A large needle such as a 16-gauge can be used with a very large syringe. The needle is inserted slightly into the lower abdomen, about an inch or so past the end of the keel, between the keel and the vent, to avoid puncturing any internal organs, and as much fluid as possible is extracted using the syringe. This fluid is straw colored and like water in consistency. Once the needle is removed it is possible that remaining fluid will drain from the abdomen. This will not cure the problem but does provide relief. Because this disease is not curable, it may be necessary to repeat the draining procedure should the affected bird swell again.

Just because a bird develops water belly does not mean that it is an immediate death sentence for it. Many birds go on to live fairly normal and relatively healthy lives apart from an occasional draining. We had a small Old English Game Bantam that developed water belly and lived with the condition for almost three years before she finally died of heart failure. We constantly monitored her swelling and only had to drain her abdomen twice in that time. Her system was able to absorb much of the fluid buildup. She would swell and then the swelling would go down. It is only when she started having trouble walking that we would have to drain her.

Just as in humans, a diet that is lower in sodium will help to control the effects of water belly. Water belly is not contagious, but the underlying cause of pulmonary hypertension does seem to have some genetic markers that can be passed to offspring. Just as if your family has a history of heart problems you would be more susceptible to it, so it seems to be the case in poultry. This is not to say that resulting offspring would ever show signs of pulmonary hypertension, but they could.

Q: Why did the chicken cross the road?
A: To prove to the possum it could be done.

Diseases

Diseases, unlike ailments, are caused by bacterial, fungal, or viral influences. There are a myriad of various diseases that afflict chickens, and it seems that every year there are new diseases, or at least a greater understanding of current diseases, new medications and vaccines, and methods of remedy or treatment. The science of treatment and prevention of diseases within chickens has, in the past, been very limited, but with a renewed sense of interest, especially in the backyard chicken, headway is slowly being made with viable treatments. What is found many times when trying to diagnose disease in chickens is that a variety of diseases will be lumped together into one general category because the symptoms are very similar. The following is a basic list of the most common diseases that you may encounter while caring for a flock of chickens.

AFLATOXICOSIS

Aflatoxicosis, unfortunately, is fairly common especially when feed is stored in large quantities. It sounds serious, but what is it? Aflatoxicosis is basically food poisoning brought on by mycotoxins in feed. These mycotoxins are highly toxic and carcinogenic and can cause significant rates of death within a flock.

Feeds that sit in a warehouse for any length of time or are exposed to moisture are the most common sources of aflatoxicosis. Any time you buy feed, make sure it is from a reputable dealer who moves a volume of feed. Check each bag as you open it to make sure that is has not been subjected to moisture. If you transfer your feed into feed barrels, be sure to use up all the feed prior to adding new. Do not purchase more feed than can be consumed in fourteen days.

Feeds that are in question should be discarded. Tainted feed can become unpalatable to the birds. Symptoms will include poor feed conversion, passage of undigested feed in their droppings, anemia, and decrease in egg production and, in severe cases, can cause lameness, muscle spasms, convulsions and death. Symptoms generally will coincide with new shipments of feed.

Because the toxins are not readily absorbed into the bird's system and are quickly passed in their droppings, recovery is fairly rapid once feed is changed. The use of bentonite, which is commonly found in food-grade diatomaceous earth, is also of great benefit as it will bind the mycotoxins and help pass them from the body.

It is very important that you manage the birds' feed closely. Make sure that there is no contaminated or moldy feed left in feeders or on the floor. Rotate and clean feed bins regularly. Do not store feed where it is susceptible to excess moisture. Watch your birds for a lack of willingness to eat the feed, and buy your feed from a reputable dealer.

AVIAN INFLUENZA

Avian influenza is one of the most dreaded poultry diseases known to man. It has the potential for creating a pandemic that can wipe out a huge percentage of the nation's birds as well as infecting humans. Though avian influenza is talked about quite often in the poultry world, it is quite rare in the United States. There are twenty-five different subtypes of avian influenza with the highly pathogenic virus being H5N1, which is heard of from time to time.

Avian influenza, or A.I., outbreaks most often occur in domestic fowl and turkeys. The source of A.I. seems to stem from direct or indirect contact with wild waterfowl and for this reason can pop up anywhere in the world at any time. Once it is contracted by domestic fowl, it spreads very rapidly by bird-to-bird contact, contaminated equipment, feed trucks, and work crews, and wild waterfowl are no longer necessary as an active mode of transmission. It is possible for A.I. to be spread through airborne transmission if the conditions are right. It has not been resolved if it is also possible for vertical transmission of the virus to eggs; it is most likely that any embryo would not survive to hatching.

The symptoms can vary widely depending on the strain of the virus. Most often, it appears suddenly within the flock and many birds just die without showing any symptoms or minimal symptoms of depression, lack of appetite, ruffled feathers, staggered walk, and deformed eggs. Sick birds will often stand in a semi-comatose state with their heads touching the ground. It is also possible to see the birds' combs and wattles turn a gray color with small hemorrhages at the tips. It would also not be uncommon for the infected birds to have profuse watery diarrhea and be excessively thirsty. Younger birds may also be affected neurologically.

But avian influenza should not be confused with other types of diseases that cause high mortality rates and similar symptoms such as Newcastle disease, infectious laryngotracheitis (ILT), acute poisoning, and fowl cholera, among others. In any large rate of death situation, it would be prudent to contact your local agricultural department to aid in the determination of the infecting disease. If it is suspected that your flock is in fact infected with a form of avian influenza, it cannot be stressed loud enough or long enough that immediate attention is required through your local agricultural department! The nation's poultry supply depends on it.

If it is determined that your birds do have avian influenza, immediate culling of infected birds is a priority, along with proper biological disposal of any carcasses, complete cleaning and disinfecting of all premises and equipment, and removal and proper disposal of any other potentially infected materials. As you can see, avian influenza is not a disease to be taken lightly. Outbreaks are rare, but they do occur and can occur at any time and in any place. If you have a large number of sudden deaths within your flock, especially if your birds are allowed to free-range near a pond or other areas frequented by wild waterfowl, it is very important to contact a person of knowledge of avian diseases immediately.

BUMBLEFOOT

Bumblefoot is a very painful infection that occurs on the pad of the foot. You will first notice your bird limping and a swelling of the pad. A dark scab will form and, if left untreated, will create a large abscess that will need to be surgically removed.

Bumblefoot can, for the most part, be prevented through simple care and well-being of your flock. Bumblefoot is caused by a scrape or cut on the pad of the foot brought on by such things as their roosts being too high, too large or square edged; the birds walking on rough ground such as sharp rocks, pieces of metal or glass; or birds that are overweight or large for their breed. Because chickens naturally scratch at the ground and walk in their own poop, bacteria get into the scrapes and cuts and start an infection.

Most bumblefoot infections are caused by the Staphylococcus aureus bacteria but can also be brought on by E. coli, Corynebacterium or Pseudomonas bacteria. These are all very quick-acting bacteria and can cause serious infection in just a matter of days.

If detected early, bumblefoot is very treatable. If left unchecked, treatment can be a long and painful experience for both your chicken and you. Prevention is your first course of action. For light birds, such as bantam varieties, keep their roosts three feet or less off the ground. For heavyweight birds, set their roosts no higher than two feet off the ground. Make sure that their roosts are free of any sharp objects such as nails, staples or splinters. Round all corners. Keep your birds off of sharp objects by raking out their pen areas to remove sharp stones, glass or other foreign objects. If your birds free-range, check that they are not scratching around in old garbage or rubbish piles. Check each bird's feet as often as possible. This is easiest done at night when the birds are roosting. A simple visual check is good but actually running your hand over each foot is better. Watch for any limping from your birds.

If your bird does develop bumblefoot, immediate care is of the greatest importance. In the early stages of bumblefoot, you may just notice a scrape or cut along with general swelling of the pad. The foot may feel hot to the touch.

You will need to get the bird onto a strong antibiotic. Amoxicillin, erythromycin or penicillin should be given per recommendations. Place the bird in a controlled area where it is on soft ground or shavings and not allowed to roost. You will need to keep the foot as clean as possible. Soak the infected foot in an Epsom salt bath twice a day until the infection is gone.

If a black scab forms on the foot, then the infection is in the intermediate stage. You will want to follow the previous instructions with the addition of removing the scab and any infectious material, apply a good antibiotic ointment and cover with a clean gauze pad and tape.

If the infection becomes acute, then you have some serious work ahead of you. The foot will be quite swollen, red and hot to the touch. There will be a large scab on the foot and a hard pus core below it. This is where it will be necessary to do surgery on the foot. This is why it is better to catch it early or do your best to prevent it from happening in the first place.

Isolate the bird and restrict unnecessary movement. Start the bird on a strong antibiotic for five to seven days prior to any surgery attempt. Soak the foot twice a day in an Epsom salt bath and clean the wound area as much as possible. Remove the scab as it softens. Massage the pus core to soften it and loosen it as much as possible. When the point of surgery comes, you can numb the area with a numbing agent such as Anbesol. Take a new, sharp scalpel and make an incision across the infected area. Keep the incision as small as possible. To control any bleeding, you can apply blood stop. Once the incision has been made you will need to work at getting the pus core out. This can be done with the aid of a dental pick. Sometimes the core is solid enough that with gentle pressure will pop right out.

Once the core has been removed, flush the wound with saline solution and apply a good coat of antibiotic ointment to the wound. Cover this with a sterile gauze pad and wrap with surgical tape. Change the dressing at least once a day and apply new antibiotic ointment with each change. If the incision is large, it may be necessary to suture the wound or use a butterfly bandage to keep the wound closed for proper healing. Keep the bird as immobilized as possible to keep it from placing unnecessary pressure on the foot. You may have to actually form some sort of cast around the foot to keep it from further damage. Though I have never tried it, some people find that cutting a tennis ball or racquetball in half and placing this around the foot and taping it in place helps protect the foot during healing. Continue the antibiotic treatment for a week after the wound has healed. Curing bumblefoot is a long process, but it can be successful. Prevention is the best medicine for bumblefoot.

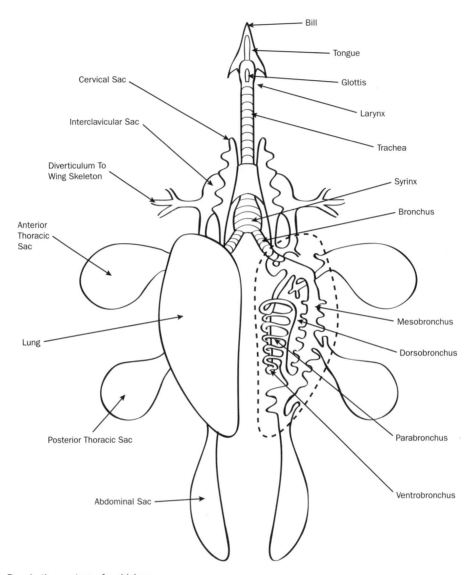

Respiration system of a chicken

Some additional alternative treatments for bumblefoot have been passed around the poultry community that include the use of Oxine AH Sanitizer and Disinfectant as a soak because of its power to kill the Staph bacteria, and good results have been discovered by using an equine thrush treatment that is normally used on horses' hooves to cure thrush. There are also some old-time remedies out there that may prove to be of some help in the treatment of bumblefoot. With any treatment, it will be best to catch the cause early and provide effective treatment to cause the least amount of harm to the bird.

CHRONIC RESPIRATORY DISEASE (CRD)

Chronic respiratory disease, or CRD as it is more frequently termed, is an extremely common problem amongst poultry flocks. The problem lies in the fact that CRD can be treated but not cured, so once a bird has it, it will always be a carrier and it can be readily transmitted to other birds within a flock.

CRD is referred to as a stress disease, meaning that it can lay dormant in a chicken's body until they are stressed for some reason and then the disease will show itself. Your birds could very well be carriers and you would not know it until some time has passed. Stress factors can be anything from a feed change to taking your bird to a show. Sudden changes in weather or a predatory attack are also two common causes of CRD showing up in a flock. It was already there, it just took a stressor to bring it out.

CRD is caused by Mycoplasma gallisepticum, or MG. There are twenty-two different strains of Mycoplasma that can affect poultry, but it is only MG that brings on CRD. There are many other respiratory infections that people mistakenly call CRD, such as a respiratory fungal infection, infectious bronchitis and airsacculitis. For accurate determination of an infection by MG, diagnostic laboratory services would be needed.

CRD can be readily transmitted by direct contact, airborne transmission and also through hatching eggs. It is very possible for you to get hatching eggs from someone and the resulting chicks would pass on MG to your flock without you knowing it until stress in your birds made it appear. Young birds are more likely to show symptoms, and it seems to be more prevalent in the winter.

CRD symptoms include coughing, gurgling, nasal discharge, frothy eyes, rattling, ruffled feathers, loss of appetite and weight loss. In laying hens, a drop in egg production would be noticed, as well as fertility and hatchability of eggs.

Mortality is generally low if treatment is administered at the first signs of infection. Secondary infections leading to infectious bronchitis or coryza can raise the mortality rate significantly.

As stated earlier, there is no cure for CRD. Survivors become immune but can still pass it to others within the flock. CRD is a slow moving disease but it takes a long time to get under control. If your birds do develop CRD, you should not sell or transfer birds or eggs to another person, as you will be passing on the disease to them.

CRD can be readily treated with tetracycline and mycin-based drugs in the early stages. Birds that show symptoms should be isolated from the rest of the flock. Treatment of the entire flock is highly recommended, but at lower dosages than those directly infected. Clean and sanitize often the living quarters along with feed and water dishes. A strong chlorine bleach solution seems to work well for this. Minimize stress to the birds such as feed changes, worming, overcrowding and adverse weather conditions.

COCCIDIOSIS

Coccidiosis, or cocci for short, is a highly contagious disease that primarily affects chicks between the ages of ten days and eight weeks. Beyond eight weeks, most birds have established a resistance to it.

Cocci thrives in damp and humid conditions. It is very important that you keep your chicks' brooder area dry at all times.

Cocci spreads through chicks very fast, so always be on the lookout for it. Chicks with cocci will tend to be listless with ruffled feathers and droopy wings. In some cases of cocci you will notice blood in their poop, but this is not always the case. They will also not want to eat and usually stand with their eyes closed. But don't confuse this with sleeping, as many chicks will sleep standing up.

This is a fast moving disease and many times the first signs you will see are dead chicks. This is one of the most common diseases that chicks get, so always be on the lookout for it to pop up. Coccidiosis is caused by the protozoa known as coccidia. Cocci is passed through the birds through their droppings. This stage is called oocyst. Oocysts can survive outside the body for up to a year and are not killed by normal disinfectant means. There are many types of coccidiosis, some of which are more severe than others.

Coccidiosis is generally an easily treatable disease with quick results if caught soon enough.

Most of today's chick starter feeds come with a coccidiostat in them, but this is by no means a fail-safe preventative. There are drugs on the market specifically designed to treat cocci. The two most common drugs are Corid and Sulmet. Corid is a 9.6 percent amprolium coccidiostat, whereas Sulmet is a 12.5 percent solution of sulfamethazine sodium as a coccidiostat. Both are good drugs for curing cocci. Many people do not like to use Sulmet as it is a much stronger drug, but we have found that it is far superior for clearing up a bad outbreak of cocci.

Both are used in the chicks' waterer at a rate of two tablespoons per gallon of water. This is to be their only source of water for a period of five days. Give this solution to all the chicks even if they are not yet showing signs, as they are most likely also infected. Make sure their bedding is clean and dry, and if on shavings, change them at least every twenty-four hours, preferably every twelve hours. If there is blood evident in their poop, make sure you change the bedding more often. It is best if they can be on a wire floor as to allow the poop to drop through and away from the chicks.

Make sure that all the chicks are drinking. If you notice any of them not wanting to drink, then you will have to give this solution by way of an eyedropper or syringe (without needle) and dribble it across the beak until it drinks. Do not force it into the chicks' beak as you run the risk of suffocating them.

Chicks seem to respond very well to this treatment and within a few days are back to normal.

Once a chick is about eight weeks of age, they have usually built up a resistance to cocci. But that is not to say that older birds cannot get it. This is especially true if you house different-aged birds together. Cocci moves much more slowly in older birds but treatment is just as effective. It is a nasty little disease but easily cured and prevented.

FOWL POX

Fowl pox is caused by a form of herpes virus. There are two main types of fowl pox that affect chickens. One is what is known as dry pox and the other is wet pox. Contrary to popular belief, they are caused by two different strains of virus. One does not cause the other, although a chicken that has one certainly can contract the other.

Both forms of pox are transmitted mainly by mosquitoes, which is why it is more prevalent in the fall and after periods of heavy rain. Pox can also be transmitted by other critters such as lice, mites and fleas. Therefore, it is very important to keep your birds free from parasites.

Dry pox forms raised lesions on the comb, wattles and about the face. These start out as raised white bumps that soon turn black. Dry pox is not fatal in and of itself. It takes about two weeks for it to run its course. Fowl pox is contagious as it can be passed from one bird to another by direct and indirect contact. It can quickly spread through your flock. There is no known cure for fowl pox. There are live vaccines that can be given to chicks or to older birds to help prevent them from contracting the disease. Once a bird has fowl pox it is best to isolate the infected bird and start a treatment of iodine applications to the lesions to help dry them up. The use of Oxine sprayed on the heads of infected birds has also proven very effective at clearing up the pox. It is also advisable to start your birds on a broad-spectrum antibiotic to help ward off possible secondary infections. Once the pox has cleared on the infected bird, it now has an immunity to further infections of pox.

Wet pox is another story. This form of pox infects the mouth and throat of the bird. As the lesions grow and spread, they can literally suffocate the bird to death. Because the lesions spread deep down the throat of the bird, there is virtually no way of treating them with iodine like you would the dry pox. Depending on the severity of the infection, the bird will stop eating and then stop drinking as the irritation becomes worse. There has been some hope given by the use of Oxine in the drinking water if administered at the first signs of infection.

Fowl pox also appears to be breed specific. If you raise turkeys together with your chickens, turkeys would not spread fowl pox to your chickens. But that is not to say that your chickens could not also be infected by transmission from a mosquito or other parasite. It just means that it will not spread by direct contact.

Fowl pox is of the same virus family as the human version of chicken pox, but you cannot catch chicken pox from a chicken that has fowl pox. But you can spread fowl pox between your chickens. Always handle your healthy birds first before working with your sick birds. After handling your sick birds, make sure you wash and sanitize your hands and change your clothes.

INFECTIOUS BRONCHITIS

Infectious bronchitis, or IB, is considered the most contagious respiratory disease of poultry. Although all ages of birds can be infected, young chicks are highly susceptible to the disease and results in high mortality rates. IB is easily transmitted through multiple modes, including direct contact, equipment, and clothing, and it can be spread great distances through the air.

IB symptoms include coughing, sneezing, rales (rattling), watery discharge from nostrils and eyes, and some facial swelling. In chicks older than six weeks, nasal discharge is usually not evident, but it can cause severe facial swelling. In older birds and laying hens, respiratory symptoms will be present along with a decrease in egg production with misshapen eggs. Egg production usually will not return to normal after treatment.

Because the symptoms of IB are very similar to other respiratory illnesses, correct diagnosis may not be available without laboratory services. Symptoms generally last for ten to fourteen days. If illness persists beyond fourteen days, a secondary cause should be examined. There is no known treatment for infectious bronchitis, though some antibiotics may lessen symptoms. Disinfecting the premises as well as feeders and waterers while the virus is active within a flock can lessen the severity of symptoms.

IB is much like a human flu bug. It comes on fast, lasts a couple of weeks and then goes away. The young and the compromised are at greatest risk. There is no real treatment, but there are vaccines that help prevent it. It is easily spread and any within a group are likely to have it. Good sanitation, cleaning and disinfecting along with strict isolation can help get it under control.

INFECTIOUS CORYZA

There are many types of different respiratory infections that can develop within a poultry flock, but one of the most easily recognizable ones is infectious coryza.

Coryza is a bacterial disease that primarily affects the eyes and sinuses of the infected bird. It is highly contagious, and once a bird gets it, even though it may be treated, the bird will always be a carrier of the disease. It is much more common in the fall and winter months, especially in tropical and temperate climate areas.

You will first notice your bird sneezing and a foamy discharge from the eyes. This is followed by a sticky discharge from the nostrils. As the infection becomes more advanced, the bird will take on the putrid smell of infection that will be very noticeable even from a great distance. The eyes and face will appear very swollen. Feed and water intake will slow or stop, and once that happens the comb and wattles may begin to turn bluish in color.

Coryza is treatable but not curable. If caught in the early stages, a strong antibiotic will usually clear up the symptoms. The infected bird needs to be immediately removed from the flock and put in a quarantine area well away from the other birds. Culling of the bird is the preferable method of eradication of this disease, as any infected bird will remain a carrier. But this is not always possible, so treatment becomes the next viable alternative.

A strong antibiotic such as amoxicillin is recommended for early signs and symptoms. This is generally given as an additive to the drinking water. The symptoms may go away, but know that the bird is and always will be infected. Coryza is a fast-acting disease and symptoms will progress rapidly. Unless you spend a great deal of time with your birds, you probably will not notice the symptoms until they reach the putrid smelling stage. At this point, it is nearly too late to do anything about it. As a last-ditch effort to save the bird, ½ cc injections of Tylan 50 are given intramuscularly in the breast muscle every other day for three doses, alternating injection sites. The symptoms will generally begin to clear within twenty-four hours of the first dose. If the bird does in fact survive, it will be a carrier for life and you will not be able to sell any birds from your flock, as you will be transmitting the disease to someone else's farm. It is highly advisable to cull out any birds that are showing any symptoms.

MAREK'S DISEASE

It's that dreaded word that sends shivers through the body of most new poultry growers. Marek's disease has been given such a stigma over the years that it is probably the most recognizable name in poultry diseases. People are frantic about how to prevent it and what can be done to keep it away from their flock. Truth is, folks, that if your chickens are breathing, they have been exposed to Marek's, which is a disease very closely related to lymphoid leukosis.

Marek's is most commonly spread through microscopic feather dander that can be carried great distances on the wind. Dust carried on your clothing can also transmit the disease from one farm to another. Marek's is so common that it is said that it kills more birds than any other disease.

But if it is so common, what can be done to prevent it?

Let's first look at what the disease really is. Marek's is from the herpes family of viruses. The most common versions of the disease are eye, visceral (tumors) and nerve. The most recognizable of these three would be the nerve version, which causes paralysis in the wings, legs or neck. Because of this, Marek's is also known as range paralysis. Some forms of paralysis may be transient, meaning that it may disappear within a few days. Without access to veterinarian services for poultry, Marek's will many times go misdiagnosed or undiagnosed. The most common home diagnosis of Marek's would be paralysis of the legs, which is visually confirmed by one leg jutting out to the front and one leg jutting out to the back and the bird unable to walk due to the paralysis.

The eye version of the disease can be detected by irregular shaped pupils, cloudy eyes (gray eye), or sensitivity to light. Marek's can result in blindness.

The visceral version, which is probably the most widely undiagnosed, should be considered if a bird is just wasting away for no apparent reason. The bird will show depression and emaciation or may show no signs at all. A post-mortem examination would be required to determine if this was in fact the cause of death. In a post-mortem examination, small, soft gray tumors would be evident in the ovary, proventriculus intestine, and muscles, and sometimes in the lungs, kidneys, heart, liver and other tissues. But if you find that you have a bird that is wasting away and no other symptoms are evident and no treatment you try seems to help, then you could suspect the visceral version of Marek's is to blame.

Marek's usually affects birds from five to twenty-five weeks of age. It can appear in older birds, but that is rare. If the bird is over twenty-four weeks old, it would be more suspect that the bird would be suffering from lymphoid leukosis, but lymphoid leukosis does not produce paralysis. Both diseases will produce internal lesions that would only be seen by post-mortem examination. Marek's disease will also cause external lesions that are noticeable on the skin, eyes and neural areas. The most common clinical signs would be lameness (paralysis), lack of coordination, paleness, unthriftiness (slow to grow or put on weight), weakness, labored breathing, and enlarged feather follicles. Lymphoid leukosis can potentially mimic most of these symptoms other than the paralysis and, therefore, is easily misdiagnosed as Marek's disease.

Marek's disease is extremely contagious but does not spread vertically to the egg. As stated before, it spreads by way of feather dander that is inhaled by other chickens, entering by way of the respiratory tract. Since the exposure to Marek's is so difficult to control, especially if you take your birds to poultry shows, the best course is prevention. Your brooders should be completely disinfected and sanitized between sets of chicks. You should also regularly clean, disinfect and sanitize your coops and pens. You can also consider vaccinating your chicks against Marek's disease.

Marek's vaccine is a live virus vaccine, which is administered to day-old chicks by way of injection at the back of the neck just under the skin. If you choose not to vaccinate your chicks, which many people choose a natural route when raising their birds, then a chick should develop a natural immunity to the disease by the time it is five months old. This is one of the reasons it is important to raise new chicks separate from your older birds. Older birds that have contracted Marek's and have managed to survive will always be carriers, and if you bring new birds into your flock, be aware that they can potentially be carriers. Once a bird develops symptoms of Marek's, there is no known cure or path of treatment, yet mortality is relatively low at just 10 to 15 percent, although the infected birds will remain carriers. Destruction of the infected bird is highly recommended.

NEWCASTLE DISEASE

Newcastle disease is one of those that you never want to happen on your farm. Fortunately, Newcastle is quite rare in the United States due to the Agricultural Department's close monitoring of poultry flocks, but it does happen. Newcastle disease (NDV) is also known as Exotic Newcastle Disease (END). The most recent outbreak of Exotic Newcastle Disease was in California in 2002 and resulted in the destruction of more than three million birds to get it under control. Prior to that, the last major outbreak was in 1971, also in California, and nearly 12 million birds had to be destroyed to get the disease under control.

There are three classifications to strains of Newcastle disease: lentogenic, mesogenic and velogenic. Lentogenic is the least deadly and is present in the United States. It causes symptoms such as coughing, gasping, sneezing and rales (rattling) with usually no mortality. Symptoms often mimic those of other respiratory illnesses. Mesogenic is a bit more drastic in the sense that it will cause severe respiratory distress along with various neurological symptoms. Mortality rate is still fairly low. Velogenic strains are deadly and, therefore, are the ones with most consideration.

With velogenic strains, it's possible for a whole flock to die before symptoms are even recognized. Although symptoms can vary, the most common would be lethargy, loss of appetite, ruffled feathers, tremors, progressive paralysis of wings and legs, and twisted neck. Once contracted, various strains have an incubation period of two to fifteen days.

The worst strains of END are brought into the United States through smuggling of exotic species of birds, eggs and non-native breeds of chickens. END can also be brought in by wild fowl, especially sea birds and pigeons. The virus can be spread through inhalation of respiratory secretions or by ingestion of contaminated fecal matter. Because some species of birds are far less affected by the viruses than chickens are, it is possible for a wild bird to shed the virus for up to a year.

Diagnosis of END is very difficult in the field due to lack of symptoms but should be considered if there is a high rate of sudden death within a flock. With any high rate of death, the local agricultural department should be quickly notified so that the birds may be tested and exact diagnosis be made.

To help control possible outbreaks of NDV or END, no birds or eggs should be imported into the United States without proper quarantine procedures. Domestic fowl should not be allowed to come into contact with wild fowl or their feces. Standard bio-security measures should be followed at all times.

RESPIRATORY FUNGAL INFECTIONS

In recent years, this has been especially tough on poultry. With extensive early season flooding followed by high heat lasting for long periods of time, many people are experiencing illness in their flocks.

At first glance this illness appears to be mycoplasma related with the respiratory issues. But this is also causing unexplained deaths of the birds, many of which do not show any symptoms. What we have found is that this disease is in fact not caused by a bacterial or viral issue, but instead a fungal problem, making it unresponsive to antibiotics.

During times of adverse weather conditions where we have high heat accompanied with high humidity, heavy rains followed by dry spells, the conditions become ripe for mold to grow. The mold is not necessarily visible. In wet conditions, such as wet litter from rain or from trying to keep your birds cool, mold will form in the litter and on surfaces. As conditions dry out, the mold becomes spores that can be carried in the wind where they are breathed in by the birds and settle into the moist areas of the lungs and trachea where they once again grow, resulting in a respiratory fungal infection.

This type of respiratory infection is not contagious, but multiple birds can develop it at one time. Many people would first suspect CRD (chronic respiratory disease) as the cause of their birds' illness as first signs closely mimic those of CRD. Respiratory fungal infections will cause pneumonia-like symptoms with deep gurgling breathing along with occasional sneezing, chirping, sniffling, coughing and gaping (stretching out neck and appearing to be gasping for air). But many times there will be no noticeable symptoms. Your birds may be up running around normally during the day and you would not notice a problem until it is almost too late.

The first response to these kinds of symptoms is often to put the bird or birds on antibiotics only to find that the antibiotics are of no use. They may offer temporary relief but the symptoms soon return. But if left untreated, this infection will continue to worsen and the bird will quit eating and drinking and soon will die. This infection is worse in young birds and birds with compromised immune systems. Most healthy birds can naturally fight off the infection unless the spore levels become too high.

The best way to detect the possibility of a respiratory fungal infection is to go into your roosting area at night when the birds are at rest. Listen closely to your birds breathing. If a fungal infection is present, you will hear gurgling in their breathing. If you have a few birds, you can pick each one off the roost and place your ear up to their back and hear gurgling from their lungs. If they are heavily infected, then the gurgling sound will be very pronounced and can be heard from a distance. With a respiratory fungal infection, you should not see any outward signs such as runny nose or watery eyes. These symptoms would be caused by other respiratory issues; but as stated above, you could witness coughing, sneezing and gaping.

Antibiotics such as oxytetracycline will only give temporary relief of symptoms from respiratory fungal infections. It will not cure it, but can possibly relieve symptoms enough to get by until you can properly treat the bird.

The best and most common treatment for this infection has come from the use of Oxine AH. Oxine is a liquid that is mixed with water to a diluted state and can be used in various ways to combat and control different bacterial, viral and fungal problems. Oxine is the only FDA and EPA product certified for direct spray over your flock. It is OMRI certified for organic use. Once mixed per the package directions, you can either mist it over your birds or, as many people do, put it in a cool mist vaporizer and fog your coop for ten to fifteen minutes each night while the birds are inside. Oxine is a contact product, which is to say that it will kill bacterial, viral and fungal diseases on contact. This is why it is important that your birds actually inhale the mist. Once inhaled, Oxine will kill the fungal spores that have settled into the lungs and trachea of your birds.

It is advised that your pen areas also be disinfected with Oxine. This can be done during the day while the birds are out, or if you have caged birds, they can be removed for a short period while you spray down their pens and then return them after the Oxine has dried. This will help kill residual mold spores before they can re-infect your birds. Oxine can also be mixed into the drinking water as a maintenance product to help kill off other internal diseases.

Respiratory fungal infections are nothing to mess with. It is important that you get your birds treatment as soon as possible to avoid serious consequences such as death of your birds. If caught early enough, it is easily treated.

Keep your pen and coop areas as dry as possible. Wet conditions are never a good thing for poultry. Disinfect regularly and listen to your birds.

SINUSITIS

As much research as I have done, I have yet to find an exact answer to what this is called. Some say that it is a form of conjunctivitis; some say sinusitis and others are just plain stumped. It is a severe swelling of one eye that results in a large hard puss ball that causes partial or total blindness in that eye. It comes on rather suddenly and the eye can swell to incredible size in mere days.

It is not contagious, though it is very debilitating for the chicken involved. There is no swelling of the head and there is only a very minimal smell associated with the infection. It only infects the one eye, which could be either the right or the left. At first glance, this affliction does not appear to be sinus related as there are no underlying symptoms. But, in fact, this is sinus related.

What happens is that a piece of food or other foreign matter gets stuck in the sinus cavity inside the bird's beak. This causes the fluid in that sinus cavity to not be allowed to drain and, therefore, backs up into what is known as the infraorbital sinus. This fluid then begins to become infected and harden. The resulting hardened mass consumes the area of the eye. Depending on the severity of the infection, this will cause partial or total blindness in the infected eye and can cause the beak to become crossed.

Because it takes away the sight of one eye, the chicken looses its depth of field, making it very difficult for it to eat or drink. This can result in death to the bird due to starvation or dehydration if left untreated.

We have found that simple treatment is effective in controlling the problem. First, the infectious core must be removed from the eye area. The bird should be quarantined to a separate wire cage to help prevent foreign matter from entering the eye once the core is removed, and for ease of treatment. A warm, wet washrag held to the eye area helps to loosen the skin around the infection. This may have to be repeated a few times a day over a number of days to get the core to loosen. When you are ready to do the procedure, dampen the core with warm water and administer slight pressure to the upper and lower lids to stretch them around the core. This will allow the core to pop out with minimal effort.

Once the infectious core has been removed, it is also possible and recommended, to use a wet cotton swab to gently press up against the sinus cavity inside the bird's beak to help force out any foreign matter that may remain lodged. This will help restore normal function of the sinuses and allow them to drain.

After the core is removed and the area gently cleaned, spread a thin layer of Terramycin ophthalmic ointment over the eye. A strong antibiotic, in this case amoxicillin, should be given at a dose rate of 250 mg twice a day for three days, followed by one dose a day for two days. The amoxicillin comes in capsule form that can be opened and the contents mixed with a favorite treat. The Terramycin is applied twice a day for five days as well. After five days, the eye should be mostly healed and actually start to seal shut, resulting in permanent blindness in the one eye. The affected bird soon will learn to eat and drink normally and adjust to life with limited sight.

Fast response seems to be the best medicine for this ailment. The faster you can treat the infection, the less chance of total blindness will result. Just because your bird develops an infection of this sort does not mean that it is an automatic death sentence. With proper care, the bird can go on to live a fairly normal and productive life amongst the flock.

Did you know?
Healthy chickens will aid their sick flock mates. In most instances, a chicken that is afflicted by blindness will be aided by others to help find food and water by gently nudging them in the right direction and through the use of vocal sounds.

Other Problems

Aside from the common ailments and diseases, there are some basic problems that are not as common, but the potential for your birds to be affected is still there. These are neither ailments nor diseases, but instead a few things that can be readily prevented with basic control over your flock. Chickens are very curious creatures and can get themselves into trouble quite rapidly if left to their own devices. None of these problems needs to arise, but on occasion they will due to a momentary lapse of judgment on our part.

BOTULISM

Botulism is not a very common problem unless your birds are kept in bad conditions or if they are free-ranged on open land. Botulism is caused by toxins from decaying animal or vegetable matter and can also be caused by your birds feeding on fly maggots which have fed on such matter. This becomes the case if dead animals or birds are allowed to decay openly or if moldy foods are given to your birds. This can inadvertently happen if you toss all your scraps onto the compost heap and your birds are allowed access. Or if you toss out old bread to your birds as a treat and there happens to be mold on the bread. It is very important that your birds are never given anything with mold on it.

Botulism is very fast acting and will result in death of the bird within twenty-four hours if not promptly treated. Botulism is marked by a sudden onset of body weakness, an inability to walk and progressive paralysis followed by death. Birds are usually seen lying with their necks stretched out in front of them. Symptoms show up within a few hours of ingestion of material. Fast action on your part is paramount if you are going to have any chance of saving your bird.

Once it has been determined that your bird is in fact suffering from botulism poisoning, separate the bird into a quiet area. Mix up one tablespoon of Epsom salts into one cup of warm water until dissolved. Using a plastic syringe, get as much of the Epsom salt solution as possible down the bird's throat. Be careful when doing this so that the solution does not go down the windpipe and drown the bird. While the bird is resting, clean up and dispose of the contaminated material to prevent other birds from becoming ill. Continue to give this solution as often as possible for three days as the only source of water. As the bird starts to recover, provide a probiotic wet mash feed by mixing two quarts of dry feed crumbles mixed with four quarts of milk (this can be fresh, sour or buttermilk). Store the unused portion of feed in the refrigerator until used up. Recovery will be determined by amount of contaminated material consumed and by speed in which treatment was administered.

Botulism is a serious and deadly ailment but can be easily prevented by making sure any dead carcasses are cleaned up and disposed of promptly and properly and by making sure that no moldy foods are given to the birds. Fly control around your pens is also very important and can be accomplished by good sanitary practices.

You know you're addicted to chickens when . . . your spouse bans you from the feed store until chick season is over.

ETHYLENE GLYCOL POISONING

Ethylene glycol, the main ingredient in most common antifreezes, is a common killer of dogs and cats, but it can also kill your chickens. Antifreeze is commonly used in automobiles as well as in solar water systems and other industrial applications. This is especially true in northern climates where it is used to keep various water systems from freezing, and in the south to help prevent boil over. If your chickens are free-ranged or are housed close to areas where mechanical work is performed, the chance of accidental ingestion of ethylene glycol is very possible.

Chickens are very curious, and if they see liquid dripping from a vehicle or a liquid in some sort of bowl or pail, you can just about bet that they are going to check it out and probably drink it. Ethylene glycol is very sweet to the taste. It would not be much different than you putting sugar in their water to entice them to drink. And chickens are quick, especially if they think you have a treat for them.

Let's say that you need to change a radiator hose on your farm tractor. The chickens are off in the distance free ranging. You climb under your tractor and drain the radiator fluid into a pan so you can replace the hose. Not really thinking about it, you push the pan aside to give yourself more room to work. The chickens, out of curiosity, come to see what you are doing. It is warm outside and seeing the fluid in the pan, they all take a turn drinking. Mmm, this is good—and before you even have a chance to notice, they have all drunk some of the fluid and go back to free ranging. A short time later, after you have changed the hose and refilled the radiator, you notice that your chickens are not moving around much. At least not like they usually do. Pretty soon you notice one or two with a strange twitch. This progresses into convulsions, and looking around you notice a couple of dead birds. By this point you had better start thinking ethylene glycol poisoning.

Unfortunately, there is not much you can do once the damage is done. It takes as little as 6.7ml per kg of weight to kill a chicken. The sudden onset of symptoms consists of depression, muscle spasms, convulsions and finally death. The severity of these symptoms would be directly related to the amount of ethylene glycol that was ingested. There is no known treatment for poultry.

IMPACTED CROP AND FLUSH

There may come a time when you find your bird suffering from an impacted crop. This is generally caused by the bird eating long strands of grass, which would be common for free-range birds. Due to chickens not having teeth to chew up their food, they will instead swallow whole blades of grass, hay or straw. These will get lodged in their crop and cause the crop to become impacted.

The crop is located near the front base of the neck. When the bird eats, the crop expands. As the food is slowly passed into the digestive tract, the swelling will go down. This is very evident just prior to the birds going to roost for the night as they gorge themselves on feed to sustain them during the night.

The easiest way to tell if a bird has an impacted crop would be to view your birds first thing in the morning before they have a chance to eat. If you have a suspected bird with impacted crop, then its crop will still be swollen and firm to the touch. If this is the case, then the bird will have to have its crop flushed or it will slowly die from lack of food.

Flushing a crop is a fairly easy procedure. There are a few different remedies out there, but this one is probably the easiest and most effective. This procedure is also easier with two people—one to hold the bird and the other to do the actual procedure.

I like using a quart Mason jar with lid for mixing but you can use whatever you have around. You will also need a child's ear bulb syringe (available at most drug stores), a crop syringe, or a drenching syringe. I prefer the child's ear bulb syringe.

1. To 1 pint (16 oz) warm water (water should be just above your body temperature), add ½ cup of baking soda and mix vigorously.
2. For a large fowl bird, fill a child's ear bulb syringe with the baking soda water. If treating a bantam-size bird, then only fill the syringe about half full.
3. Place the bird standing on a table in front of you. If you have a helper, have them hold the bird standing so that you may easily work at the front of the bird.
4. Hold the bird's head back and open its beak. Look to the back of the mouth and you will see an open hole and one that opens and shuts. The one that is always open is the throat. The other is the windpipe. If you get water in the windpipe then you will aspirate the bird and it will drown.
5. Now take the filled syringe and put it in the back of the throat (the open hole) of the chicken and slowly expel the baking soda water solution down its throat. Set the syringe aside.
6. Gently massage the crop area to help loosen the mass and then take your hand and bring it up tightly against the base of the crop of the chicken. Gently work the mass upward along the throat and out the beak. Chickens do not have the ability to vomit and so you are in essence doing it for them.
7. Do not turn the chicken upside down.
8. Let the bird rest for a few moments, and then repeat the whole procedure at least two more times, allowing the bird to rest a few moments in between each flushing.
9. Now, place the bird in a cage or pen by itself with only water available. Do not give the bird anything to eat for the next 24 hours. Yes, she is going to protest, but you want her crop to return to normal before giving her anything.

10. The following day, after the 24 hours is up, crush 1 tablet of Selenium and mix with 1,000 mg of vitamin E. I prefer the liquid gel caps; I just cut the tip off and squirt out the contents. Take 1 slice of bread, cut up, and mix in enough milk to make a wet mush. Add the Selenium and vitamin E and mix. Give this mixture twice a day for a week. To the birds drinking water you can add 1 tablespoon of Apple Cider Vinegar for each quart of water.
11. After 4 or 5 days you can start introducing small amounts of unsweetened applesauce or anything that has no bulk to it. Baby foods work great for this time period. Also a bit of cod liver oil or olive oil can be beneficial to help things keep moving.
12. By the 7th or 8th day you can introduce small amounts of crumble style chicken feed. Do not feed her pelleted feed or any cracked or whole grains for about the first 2 weeks after the flushing.

After a couple of weeks, she should be pretty much back to normal and eating normal foods. Keep a watch on the bird, checking the crop every couple of days to make sure there is not a relapse.

As stated before, birds that free-range are usually more susceptible to getting impacted crop, as are those who are given large clumps of weeds while you are working in the yard. To help combat this possible problem, it is best to let your birds free-range over mowed fields. If you want to feed your birds weeds from your yard, pile them up and then run over them with your lawn mower to cut them up into tiny bits before giving them to your birds. Your birds will benefit from the added greens, but a moment of time spent cutting them up will save you and your bird a week of anguish.

Did you know?

Chickens can swim. Though they do not have the feather oils that waterfowl do, they still have the ability to float for short periods of time. In fact, some chickens enjoy bath time so much that they will literally fall asleep in the warm water and their heads must be held up.

WHITE BIRDS TURNING YELLOW

One day you walk out to your poultry pens and notice that your nice White Leghorns or your Delawares are now a dingy shade of yellow. Do the have j aundice or some other strange disease? No, really they are fine.

There are two main reasons for your white birds turning yellow. The first and most common reason is that you are feeding them too much corn. The other reason is too much sun. These are all too common problems when you keep white birds.

A diet high in corn is not good for white birds. Many people supplement the birds' regular feed with cracked corn or scratch. The corn reacts with the coloring of the feathers and turns them yellow. By cutting back on the corn intake, the feathers will gradually turn back to white.

By the same token, if you have white birds and they are exposed to too much sunlight, this can also cause their feathers to turn yellow. Combine these two things, such as for those folks living in the South, and you have Leghorns that looks more like Buff Orpingtons.

Some simple precautions can be taken if you are going to be raising white birds. First, keep the consumption of corn to a minimum. Secondly, provide plenty of shaded areas for your white birds to get out of the direct sun.

Yellowing is of greatest concern for those who show their birds. You really don't want to take your potential prize-winning Delaware to the local fair looking all dingy and yellowed. If you do show birds, then along with the bathing process, it is possible to use a bit of bluing agent to take the yellow out of the feathers. But be careful, as you might go the other way and actually dye the feathers a pretty shade of blue. Sometimes the yellowing gets so bad that the only way to get rid of it is to wait for the bird to molt.

Yellowing is not detrimental to the health of the bird; it is only detrimental to the look of the bird. So don't let a yellow bird make you blue, just realize that there are things that you can do to help prevent it.

CANNIBALISM

If your birds are suffering and dying from cannibalism, you have some serious problems within your flock. Cannibalism should never happen in any flock, especially not in a backyard flock.

Granted, the chicken is the nearest living descendant of the Tyrannosaurus rex, which was a fierce meat eater. And yes, it is okay to feed your chickens meat, which, contrary to popular belief, does not turn them into cannibals. Cannibalism within a small flock is brought on by overcrowding, boredom and unchecked feather picking.

The greatest instance of cannibalism happens in the large commercial egg-producing plants where the hens are kept cramped in small battery cages with nothing to do but eat, poop and lay eggs. Cannibalism is so rampant under these conditions that the growers automatically clip the top beaks of the hens to help prevent them from literally eating each other to death. These hens are so tightly packed into the cages that most of the time they cannot even turn around. I think they try to eat each other just to create more room.

The same thing will take place in your backyard flock if your birds are kept in overcrowded conditions. Your birds should have as much freedom of movement as possible. The more the merrier. If you can't provide your birds with adequate room to roam, then keep fewer chickens. Each bird should be allowed at least four square feet of space as a minimum. Less space than this and your birds could be considered overcrowded.

If chickens are overcrowded, they also will suffer from boredom. Chickens will take on some strange activities if they are bored. The worst would be feather picking. Chickens are active, social animals that need constant stimulation to keep them from becoming bored. Free-range chickens have a constant supply of stimuli to keep them occupied. Penned birds have less stimuli, depending on the size of cage they are kept in. Caged birds have very little stimuli available and, therefore, turn to deviant behavior to keep their minds working.

When birds get bored, the first course of deviant behavior generally is feather picking. This is where one bird will pick the tail and back feathers from another bird. This action needs to be stopped as soon as it is noticed by separating the offending bird or birds. The bird whose feathers are being picked will eventually start to bleed at the site of the picking. Once the offending bird gets its first taste of blood, there is little to nothing that you can do to break the habit. A bird that has tasted blood will continue to want blood, which leads to cannibalism.

This leads to an interesting question that we are asked many times: Is it okay to feed meat to your chickens? The simple answer is, yes. And if you do feed your chickens meat, will it lead to cannibalism? No. I would not suggest feeding your chickens raw meat as this can lead to an increased chance of worms. But cooked meat is perfectly fine. Our chickens get all of our meal scraps including pork, beef, turkey carcasses and yes, even chicken. But it has been properly cooked and is usually in bite-size pieces and it has never led to cannibalism.

Cannibalism should never happen on any farm. If your birds are housed in large enough quarters, and always the bigger the better, they should not pick at each other. If your birds are kept occupied with things to jump on and off of, hide under or to play with, such as a hanging head of lettuce or cabbage, then they should not get bored. If feather picking is kept in check by way of separating the offending bird and also by the use of anti-pick sprays, and if your birds are fed a nutritional diet, then cannibalism should never be a concern.

If by some rare chance you do experience a case of cannibalism, take a look at the conditions the birds are being kept under and make the necessary changes. Separate the offending birds; it is usually to your best advantage to cull them out, as the act of cannibalism will continue. Just as once a chicken determines that a certain treat is good and goes crazy every time that treat is given, a chicken that turns cannibalistic realizes that another chicken means treat and they will kill it to eat it as a treat. Cannibalistic behavior cannot be broken. Do not clip the chicken's beak, as this is a cruel practice that usually ends up maiming the chicken. Chickens do not turn cannibalistic just because they want to eat another chicken; they turn cannibalistic because of something we are doing wrong in their care.

Medications

The following is a list of medications, compiled from various sources, of products commonly used in the treatment of various ailments found in poultry. This is by no means a complete list, nor have any of these dosages been confirmed with a veterinarian. As with any medication, it is recommended that you speak to a trained avian veterinarian before administering treatment.

Most medications used in the treatment of poultry ailments are designed for and dedicated to the treatment of other species of animals and, therefore, the dosage recommendations must be reconfigured for use on poultry. Most medications are used in an off-label situation; therefore, it is by your own judgment that these products be administered.

Be aware that there is a potential for life-threatening reactions to occur by the use of various medications. As with all species of animals, there is a potential for allergic reactions to take place. There is also the potential for incorrect administration of medications that can result in the death of the bird.

Also be aware that the use of medications on your poultry results in the possible contamination of the eggs and meat of the bird. Therefore, it is necessary to speak with a trained avian veterinarian to determine withdrawal times for during and after treatment.

The following list is merely a guideline. If you are to administer your own treatment without first speaking to a trained avian veterinarian, then you use these guidelines at your own risk and the risk of death to your bird.

VACCINES

Marek's Disease
0.2 ml subcutaneously at one day old to prevent range paralysis (not useful after one day of age)

Newcastle-Bronchitis
One day old or older, mix in drinking water or intranasal/intraocular, effective for ninety days. To vaccinate intranasally, place finger over one of the bird's nostrils and place one drop of vaccine in the other nostril. Do not release the bird until the vaccine has been inhaled. To vaccinate intraocularly, place one drop of vaccine in the eye.

Poxine-Fowl Pox
Vaccinate with wing web applicator anytime after six weeks of age to prevent fowl pox, annual booster. Follow directions supplied with vaccine.

INJECTABLES AND INGESTIBLES

Ivomec
Wormer, ⅛ to ¼ cc orally for worms; does not kill tapeworms, does kill some lice and mites

LA-200
Broad spectrum antibiotic, 1 cc orally and 1 cc intramuscularly twice daily for 5–7 days

Omnimycin
Chronic respiratory disease antibiotic, ½ cc for 3–5 days, given orally

Penicillin
Antibiotic, ½ cc for 3–5 days for wound infection; use 3 cc orally for 3–5 days for cholera

Tylan 50
Antibiotic, control of chronic respiratory disease and infectious coryza, ½ cc intramuscularly every other day for 3 doses alternating injection sites

Valbazen
Wormer, ⅛–¼ cc orally for tapeworms

Vitamins A and D
¼ cc intramuscularly once every 2–3 weeks in brood stock to improve fertility and hatchability

Vitamin B-Complex
½ cc intramuscularly daily to increase appetite and energy Vitamin B-12
½ cc intramuscularly daily to increase appetite and energy

TABLETS AND CAPSULES

Fish Cillin
Ampicillin, antibiotic, 1 capsule per day for 5 days

Fish Cycline
Tetracycline, antibiotic, 1 capsule per day for 5 days Fish Flex
Cephalexin, antibiotic, 1 capsule per day for 5 days given with food to treat wound infections and other infections sensitive to penicillin

Fish Mox
Amoxicillin, antibiotic, 1 capsule per day for 5 days given with food to treat wound infections and other infections sensitive to penicillin

Fish Zole
1 tablet per day to help treat canker

Wormazole
Wormer, 1 tablet, then repeat in 10 days for treatment of round, cecal and tapeworms

Vitamin B-12 100mg,
1 tablet daily to increase appetite and energy

Cod Liver Oil Capsule
1 tablet 2 to 3 times per week, great source of vitamins A and D

SOLUBLES

Agrimycin
Antibiotic, treatment or control of fowl cholera, chronic respiratory disease and infectious synovitis, 1 teaspoon per gallon of water for 10–14 days

Amprol 9.6% Oral Solution
Coccidiostat, for the treatment of coccidiosis, 1 teaspoon per gallon of water for 3 days

Aureomycin
Antibiotic, treatment or control of fowl cholera, chronic respiratory disease and infectious synovitis, ½–2 teaspoon per gallon of water for 7–14 days

Bacitracin
Antibiotic, 1 teaspoon per gallon of water

Corid 9.6% Oral Solution
Coccidiostat, for the treatment of coccidiosis, 1 teaspoon per gallon of water for 3 days

Duramycin-10
Antibiotic, control of chronic respiratory disease, 2 teaspoons per gallon of water for 7–14 days

Gallimycin
Antibiotic, control of chronic respiratory disease and infectious coryza, ½–2 teaspoons per gallon of water for 10–14 days

L-S 50
For the use in chicks up to seven days of age for the control of airsacculitis. 1 teaspoon per gallon of water for first 7 days

Neomycin
Antibiotic, treatment of chronic respiratory disease, ½ teaspoon per gallon of water for 5 to 7 days

Panacur/Safeguard Paste
Wormer, 1 BB-size piece given orally for the treatment of worms

Sulfadimethoxine Soluble
Use for the treatment of disease outbreaks of coccidiosis, fowl cholera, and infectious coryza, ½ fluid ounce per gallon of water for 5 days; mix fresh solution daily

Sulmet
Coccidiostat, treatment of coccidiosis, also control of infectious coryza, 2 tablespoon per gallon of water for 3–5 days

Terra-Vet 10
Antibiotic, control of chronic respiratory disease and infectious synovitis, 1 tablespoon per gallon of water for 10–14 days

Tetroxy HCA-280
Respiratory antibiotic, control of infectious synovitis, control of chronic respiratory disease and air sac infections, control of fowl cholera, treatment level: 200 mg - ⅛ teaspoon per gallon of water; 400 mg - ¼ teaspoon per gallon of water, 800 mg - ½ teaspoon per gallon of water for 7–14 days

Tylan Soluble/Tylosin Tartrate
Respiratory antibiotic, control of chronic respiratory disease, 1 teaspoon per gallon of water for 7–10 days

Wazine 17/Piperazine
General wormer, 1 fluid ounce per gallon of water for twenty-four hours for birds over six weeks of age as a general first round wormer, kills round worms and nodular worms

Wazine 34/Piperazine
General wormer, ½ fluid ounce per gallon of water for twenty-four hours for birds over six weeks of age as a general first round wormer, kills round worms and nodular worms

Zimecterin Paste 187
Wormer, 1 BB-size piece given orally for the treatment of worms

OTHER MEDICINES

Ear Mite Medication
Apply several drops per ear twice daily for 2–3 days

FiproGuard
Antiparasitic, for treatment of external parasites. 2 drops placed on skin at back base of neck; for treatment of stick-tight fleas, 2 drops placed at back base of comb, repeat in 2 days

Gatorade

Used as a supplement to aide in the replenishment of vitamins and electrolytes during times of stress such as heat prostration. Orange seems to be the most favored.

Ivermectin Pour-On

Wormer, antiparasitic, 4 drops for bantam breeds up to 8 drops for large fowl, placed on skin at back base of neck; for the treatment of worms and external parasites such as lice and mites

Red Cell

1 tablespoon per gallon of feed

Scarlet Oil Spray

Apply liberally to legs and feet to prevent and treat scaly leg mites

VetRx

1 drop under each wing and on chest, may be added to drinking water or used as a mist per instructions

Vitamin and Electrolyte Plus

Vitamin supplement, use for vitamin deficiencies and during times of stress, 1/8 teaspoon per gallon of water

For additional drug information please visit this site:

www.drugs.com/vet/chickens-a.html

National Poultry Improvement Plan (NPIP)

Here is a wonderful program that most everybody seems terrified of. If you ask people about it, they seem reluctant to give you any information. People seem to shy away from it because it is a government program and you have to have government people come to your property to do testing on your birds. People think of it as just one more way for the government to get involved in and control our business. We, too, were scared of it until we learned the facts.

Granted, each state is different in the way they conduct their NPIP program, but it is a federal project and, therefore, should not be much different from state to state. I can only tell you about our experience and how pleasantly surprised we were.

The National Poultry Improvement Plan (NPIP), is a federal program with the sole intent of testing poultry for pullorum and typhoid. They will also test for diseases such as Marek's, exotic Newcastle and avian influenza. Trust me, you don't want any of these—and if you do have them, then you don't want the birds, and you certainly would not want to be responsible for the spread of these diseases.

The biggest fear that people have is that if the government comes out to test your birds and one or more come back positive that you will have to destroy your flock. Let me just say that if your birds have pullorum or typhoid, you will want to be destroying your flock and you will want to do it faster than the government.

Both pullorum and typhoid are extremely rare in the United States. Both diseases are usually brought in by outside sources where someone smuggles birds or eggs into the country illegally. They both spread very rapidly and have the potential for wiping out a good portion of the population of the nation's birds. I don't think that you could feel good about yourself if it was your birds that caused the next typhoid outbreak.

Prior to having our birds tested, I scoured the Internet for people who had gone through the process and tried to glean as much information as possible, which was little to none. But I also knew that if we were going to move forward, then we were going to have to be tested. To us, our birds are our pets. We have at least seventy birds that are named and many of them will come and sit with you or come in the house on occasion, and like with any pet, how could we face the fact of having to put them down if the testing did not come back in our favor. I called our State Department of Agriculture. I talked with different agriculture agents. I had different agencies send as much information as possible, which I really didn't want to do because then they had our address and I was envisioning men dressed in white lab coats peering over our fences with binoculars and notepads trying to figure out how they could destroy our birds. Boy, I couldn't have been any farther from the truth.

I finally got up the nerve to call our Department of Agriculture Poultry Specialist and tell her who I was and what we were looking to do. She was the nicest lady and reassured me that her department was not out to destroy my birds or turn me in to other authorities because we kept poultry. She assured me that her department didn't care whether or not I was even supposed to have poultry where we live. Their only concern was that the poultry that I had was healthy and clear of these diseases.

But this is the government we are talking about. They always seem to get into your business one way or another. After all, they are all interconnected right? So I was still very skeptical. She gave me the number for our local poultry inspector and wanted me to contact him directly to set up a time to come out. Oh great, now two government officials were going to have our number and address. So again I held off. I wanted to talk with a local inspector that we had seen many times at our local livestock auction. We only knew each other by face, so I figured it was safe to talk with him. I gleaned as much information from him as possible about how they test and what they look for when they are on your property. He was not an actual poultry inspector but had assisted on several testings. I still could not get a definitive answer as to what they look for or what was needed from me personally to pass the testing.

My thoughts and basic information had led me to believe that I had to have a closed flock, wash down facilities, immaculate pens, disinfectant and sanitizer at every possible area, show-ready birds, plastic gloves and booties and complete documentation for each bird, and not a sniffle or sneeze from the whole flock.

Now granted, our birds are very well cared for, but they are production birds and kept on dirt in fairly large screened pens. We have many birds that are kept in a very large common pen. Their pens are raked out at least once a week. Waterers are cleaned and disinfected at least once a week. They are fed quality feed and so on. But was that going to be enough? I am still envisioning multiple inspectors running around taking soil and air samples along with water and feed samples. I was driving myself crazy.

I got up the nerve to bite the bullet one morning and called our local inspector. I was shaking the whole time I was talking to him. He was a really nice guy but was straight to the point. He asked me how many birds we had. That was it, just how many birds we had. He set a test date for the following week. Okay, that didn't help to ease my fears. So, before he came out on that fateful day, I made sure everything was as in order as I could possibly have it. I even made a promise to all our birds that if they behaved themselves and nobody as much as sneezed, then they would get an extra helping of scratch that day.

Well, the day came with the inspector arriving at 8:00 A.M. I was out there to greet him as he drove up. It was just one inspector and he wasn't even wearing a white coat. I was still anticipating a large white van to pull in any moment with the rest of the inspectors on board, but it never happened. We exchanged pleasantries and I started in with the questions. What did he need from me? What did he need from the birds? On and on and on I went. He just kept smiling. I figured this guy had heard all this a million times before. Once I stopped rambling, he got down to his list of questions for me. He asked again how many birds we had, if there were a fairly level spot where he could set up a table and if he could get a small glass of water. That was it for his questions. Actually, he asked one more question: How many birds did I want tested? What, he isn't going to test them all?

I helped him with his table and led him to our poultry yard. He commented on how he liked our set-up. He decided that out of all our birds he would take a random sampling from each pen and also our main yard. He got his testing supplies out and we brought the first bird. A quick poke of the vein under the wing, a drop of blood to test with, a quick swab of the mouth and the first bird was done. He was testing as fast as my wife and I could bring him birds. We tested twenty-five birds out of our flock in about a half hour's time. The birds were done. I am thinking, okay, now comes the real part, all the coop inspections. He never went in the coops. He was finished except for some paperwork. He packed up his stuff and we went back out to his truck. It was after the testing that I could get him to open up a bit. The testing for pullorum and typhoid was negative. Now he would send off the swabs for testing for the other diseases. They only do a random sampling because if one bird has it then it has usually spread to other birds fairly quickly. They take a general look around just to make sure the birds are being cared for and that there are not excessively dirty or unsafe conditions. They watch for parasites as they lift the wing for testing. Other than that, they just look for general good health of the birds. It's a quick and simple process.

The inspectors will also offer suggestions of things that you could do to improve the quality of life for your birds if there is something quite obvious that you are neglecting. But they are really just there to test the birds for potentially deadly diseases that could wipe out not only your flock but also the nation's flock. I asked the inspector how often a positive result comes up. He said that in all his years of testing, he has never seen a positive result. That is not to say that there never is one, he just had never seen one. I asked him about the procedure for if they do come up with a positive result. If anything comes back positive, then they will schedule another visit and try to narrow down which bird or birds showed a positive result. From those birds, they will take further blood sampling to verify the results. If there is still a positive result, then the bird or birds in question will have to be destroyed by you or by them, preferably by them so that they can dispose of the carcass in a healthy manner. Then, retesting is done every couple of months to see if any other birds come back positive. Once you are tested clean, it goes to the standard once a year testing.

These inspectors are here for one purpose and that is to keep our nation's poultry population safe. They are not out to shut you down or turn you in for having poultry where you shouldn't. They are not out to get you. All my fears were certainly unfounded. This is a completely voluntary program and in most states it is free to have the testing done. It is well worth it to you and to all of us to have your birds tested once a year. It is quick and simple.

The advantage to this is that you have peace of mind that your birds are not going to be responsible for spreading a deadly disease. You will be able to legally ship and transport eggs, chicks and birds across state lines. You will also be able to show your birds. And you will be able to confidently and proudly tell a prospective buyer that your birds are NPIP certified.

So, put away your fears and get your birds tested. You will be happy that you did.

Appendix A: Chickens and Children

Chickens and children are synonymous with each other. What is cuter than a child holding a small ball of fuzz? Chickens can be wonderful learning tools for children. They learn how a chicken hatches, grows and goes on to hatch its own chicks. They learn responsibility with feeding, watering and general care of an animal, but they can also learn the cycle of life and death.

I have always felt that every child needs to have chickens, if for only a little while. The vast amount of information that a child can learn will follow them throughout his or her life. I can remember as a young child always having a fondness for chickens. I was never into the other fowl such as ducks and geese, and peacocks were nice to look at but they were too noisy. Turkeys were for Thanksgiving and guineas to me just looked prehistoric. My heart was for the chicken. I would send away for every chick catalog I could find and spend hours looking over the pages and making lists of the birds and supplies I wanted. I finally got my first chicken when I was seven.

A man and a lady who lived just down the street from us had a rooster and every morning I could hear it crow. I would sneak down to their house and watch his chickens roam around his barns. I always viewed this man as old and ornery, but his wife was like the stereotypical farmer's wife. She was a sweet lady and would invite me in when she would spot me sneaking a peek at their chickens. This sneaking around kept up for months, and one day the man, Ed, handed me a nice white hen. I think I was the proudest kid in the whole valley and I now had my very own chicken, but I really didn't have a clue as to what I was supposed to do with it. My parents both worked at the time and neither of them were the farming type, so I was kind of left to my own understanding of what it meant to care for a chicken. I don't remember this hen, which I named Henrietta, ever laying an egg. Later on I figured out that she was probably too old to lay and that is why Ed gave her to me, but that didn't matter because I had a chicken just the same.

Henrietta went everywhere with me. I carried her around on all my adventures. She even slept with me in my room. About the only place she didn't go was to school. But we had fun together. I could put her down outside and she would just follow me around like a little puppy. I knew she needed food and water but that was the extent of my knowledge, but I was bound and determined to learn all I could.

She lived about two more years and passed away from what I take was old age. It was a devastating time, but I did not let that deter me.

I remember when we, as a family, would head north on the freeway and we would pass by a huge white barn that sat in a little valley. It sat empty for years and I always wanted that barn so that I could raise chickens. It became a family joke that that was my barn and I was going to be a chicken farmer, but as the years passed the barn came into a bad state of disrepair and was finally torn down to make way for office buildings. I would never have that big white barn for my chicken farm, but I was still not deterred.

We moved into a house at the base of the mountains back up in the woods. I was not big on moving, but this place was great because it had its own barn and I instantly thought of chickens. So again I got all the hatchery catalogs that I could send away for and spent hours figuring out how I was going to get chickens once again. I worked hard taking care of the neighbor's yard working for two dollars an hour and saving up until I had enough to buy a set of chicks. On that fateful day I placed my order for twenty-five Rhode Island Reds and then patiently waited for a call from the post office that my chicks had arrived. In the meantime I set up a brooder area out in our old pump house and had everything as ready to go as I could figure. I still really didn't have a clue as to what I was doing, but I was learning. The chicks arrived in perfect condition and soon they were happily playing in the brooder. I would feed them in the morning before school and my time after school was spent playing with them until nighttime. Soon they were big enough to move to a pen in the barn. I built them a large indoor pen and also a large outdoor pen so that they could move back and forth. They ate and they grew and at about eighteen weeks the first hen laid her first egg. Soon others were also laying and I was proud that I was able to finally have, what seemed to me, a huge chicken farm.

They were all just starting to lay well when the dreadful day arrived when I came home from school to find the neighbor's dog had ripped open the cage inside the barn and killed every one of my chickens. Needless to say I was not happy. I had worked so hard only to have this happen. I was frustrated and almost gave up on chickens. I did the rabbit thing. I did the puppy thing and went through the car stage, but my heart was still with chickens. I continued to learn what I could about caring for and raising poultry. I just never gave it up. I was bound and determined to have myself a chicken farm. So now I keep up to 300 chickens in various breeds and am as content as can be.

Though I didn't realize it at the time, raising poultry as a small child taught me perseverance in the face of adversity. It taught me that there is something out there that relies on me for its existence and so I had better be dutiful in my care so as not to let them down. It taught me about life and death. But it also helped teach me the more basic skills such as math (counting chickens, eggs, figuring out what an order will cost and so on); reading (I was constantly studying how to raise poultry); history and geography (I was learning where different breeds came from and how); carpentry and construction (as I had to learn to build their pens and coops and how to make them strong enough to keep out predators). Chickens have taught me responsibility by making sure they are fed, watered and cleaned up after and they have taught me to look at life differently.

Chickens are great learning tools for children. Yes, there is heartache that comes with raising chickens, but then there is heartache in life. And it doesn't matter how bad a day gets or what troubles the world throws at you because when you take a walk out to the chicken pen, they are always happy to see you.

Chickens make great pets. Contrary to popular belief, they are quite smart. No, you can't teach them a myriad of cute tricks, but you can teach them some tricks. They will generally come when they are called, they will sit on your lap or shoulder and chickens love to be tickled. Big roosters can be friendly and even loving. They make diapers for chickens so they can now come in the house without making such a mess. They will chatter to you and because of you. Chickens are very social animals and need the constant contact. The more you interact with them, the more they will do to please you.

Chickens are not the nasty creatures that some people make them out to be. Chickens are only as nasty as the conditions they are kept under. If they are dirty and stinky, then it is because they are allowed to live that way. Chickens take a bath usually once a day. Yes, it is a dirt bath, but this is the way that they clean their feathers and help rid their bodies of parasites. Chickens can also be bathed in water, with shampoo and primped and pampered. Their butt feathers can be trimmed if they tend to get poopy. Their nails can be trimmed if they get too long. If you think chickens are nasty animals, go visit a poultry show and see how chickens are cared for the right way. Now, I agree, if you have 300 birds you cannot be expected to get out there and give them all a bath, but they also do not need to live in terrible conditions. This is part of the responsibility that your child or children can have by taking care of the living conditions of the birds. Chickens are great for children and they can also be great for parents as well. You can learn together as a family and find mutual enjoyment in your feathered friends. Chickens, if given the chance, can become just like any other household pet. As we chicken people say, they are our chickdren. And with us, our children are all grown and moved out on their own, so our chickens really have become our chickdren.

Appendix B: Fun Ways to Use Your Eggs

HARD-BOILING FRESH EGGS

When working with fresh eggs versus store-bought eggs, you will notice lots of differences. Yolk color is different, texture is different and taste is different. But one of the greatest differences that you will realize is that a fresh egg hard-boils differently than a store-bought egg.

If you just throw some fresh eggs in a pan and bring them up to boil and process them the way you normally would, you will end up with hard-boiled eggs that look more like scrambled eggs by the time you are finished. Not something to do just before guests arrive and you are to be treating them to deviled eggs.

Store-bought eggs sit in a cold storage warehouse for up to three months before they ever hit the store shelves. The egg producers do this for a reason. It gives the egg time to actually dehydrate so that the membranes inside the egg pull away from the shell. This allows you, as the consumer, to peel a hard-boiled egg with ease.

There are a few different methods to hard-boiling fresh eggs so that they come out clean and not all torn up. The easiest method would be to leave fresh, unwashed eggs sitting on your counter for about two weeks. This is not always possible if it is really warm or because of sanitation purposes. You could also leave washed eggs in your refrigerator for a couple of months before using them, but this is usually limited by space.

What we have found works the best is very similar to how you would normally hard-boil a store-bought egg, but with some minor adjustments. In a saucepan large enough to hold the amount of eggs you wish to hard-boil, bring enough water to boil that would cover the eggs by an inch or more. Once the water is at a rolling boil, add a good amount of salt, say two or three tablespoons. Place each egg on a spoon and gently lower them into the boiling water until all eggs are in the pan. Boil for approximately fifteen minutes, depending on your desired doneness. Remove the pan from heat and run cold water into the pan until the eggs are cold to the touch. Remove the eggs from the pan and process as usual.

BLOWING OUT EGGS

Times will arise when you find it necessary to blow out an egg or two. Whether it is for craft projects or you just want to save that special egg, blowing out eggs is easy.

Take the egg that you wish to blow out and shake it to help break up the yolk inside. Take a darning needing and carefully poke a hole in each end of the egg. You can also use a small drill bit to drill a small hole in each end of the egg. Now blow gently on the one end to force the contents out the other. As you get the egg contents to come out, occasionally run a little warm water into the egg. Cover both holes with your fingers and gently shake the egg. This will help thin out the egg contents. You may find it necessary to break up the yolk with a small wire if shaking does not do it for you. Continue blowing out the egg until empty.

Now mix up a solution of 50 percent bleach and 50 percent water. Place a piece of tape over one hole and fill the egg with this solution. Cover the other hole and gently shake the egg. Remove the tape and drain out the bleach solution and let the egg dry completely. It is now ready for decorating or using in craft projects.

NATURAL EGG DYES

Whether dyeing eggs for Easter or decorating eggs for a craft project, you can run down to your local store and get an egg coloring kit—that is if it is March—or you can be more adventurous and create your own dyes from ingredients you probably already have in your cupboards.

Creating natural dyes is limited only by your imagination. Just about anything with color can create a dye. Some things work better than others and with a bit of trial and error you can determine what works best for your situation.

The key to dyeing eggs is white vinegar. Vinegar dissolves the hard calcium of the eggshell, making it more porous so that it will better accept the dye.

Take however many eggs you will need and hard-boil them. After they are hard-boiled, place them in the refrigerator to cool completely; this usually takes a couple of hours. While the eggs are cooling you can gather up your dyeing materials. You will need a saucepan for boiling down your ingredients, some white vinegar and small bowls to put your dyes in when they are complete. Keep in mind that some natural ingredients will take just a pinch to create the color you want where others may take a handful or two.

Place the saucepan on the stove and add your ingredient. Add water to cover by about an inch. Bring to a rolling boil and continue to boil until you reach the desired color level. Add in two or three teaspoons of white vinegar and remove from heat. Strain the liquid into your bowls and allow to cool. Repeat this procedure for each desired color. Now you may take your eggs and, using a spoon, carefully lower them into the dye. Roll the eggs around in the dye to get complete coverage. When the egg has reached the desired color, remove it from the dye and blot it dry with a paper towel or napkin. Place it in an egg carton to completely dry. The eggs will have a dull finish when dry. If you desire a shinier finish, you can rub on a light coat of vegetable oil.

There are many ways in which to dress up your eggs if you so desire. You can write or draw on your eggs with different colors of crayons before placing them in the dye. The wax from the crayons will not allow the dye to penetrate the egg and, therefore, will give nice patterns once dyed. You can also take old patterned nylons and wrap them tight around the egg and tie off with a rubber band. Dye the egg as usual and let completely dry. Remove the nylon to reveal the pattern left. Use pieces of tape cut into interesting shapes and placed on the egg. Dye the egg and let dry. Remove the tape to reveal the image. This is fun if you dye the egg a light color, let dry, apply your tape and then dye a darker color. You can use rubber bands around the egg in different sizes and patterns to create a tie-dyed look. You can also create a fun textured look by blotting on dye with a sponge or crumpled up newspaper. Again, you are limited only by your imagination.

Below is a list of some of the more common natural ingredients that you could use to create your dyes:
- Blue - blueberries, red cabbage leaves, purple grape juice
- Yellow - turmeric powder
- Orange - carrots, paprika
- Pink - beets, cranberries, red grape juice
- Red - pomegranate juice, cranberries
- Green - spinach leaves
- Purple - violet blossoms, blackberries, red wine

There are a multitude of ingredients that could be used to create your dyes. Think if it will stain your skin then it will more than likely dye an egg. Use your imagination and have fun. Remember, what will stain an egg will probably also stain your work surface, so put down plenty of old newspaper to protect the surfaces.

ALTERNATIVE EGG DYES

Easter is not the only time of year when you might find yourself needing to color some eggs. Unfortunately, about the only time of year that you will find egg-coloring kits in your local store will be in March and April. But what if it is September and you are wanting to make a wreath out of colored eggs for the fall season? Have no fear; there are other great ways to color eggs. Besides the natural dyes listed in an earlier section, there are other ways to get amazing colored eggs.

Kool-Aid drink mix makes a great dye for eggs. Take a packet of unsweetened Kool-Aid and mix it with two-thirds cup water. Kool-Aid contains citric acid, so it does not need the addition of white vinegar. If you find that the Kool-Aid mixture itself does not do a satisfactory job of dyeing the eggs, then add two teaspoons of white vinegar to the mix, but you should not have to, as Kool-Aid seems to do just fine on its own.

Here is a list of some basic colors:

- Cherry - red
- Berry Blue - blue
- Strawberry – red-orange
- Orange - orange
- Grape - brownish purple
- Lemon Lime - green

Another alternate method of dyeing eggs is by the use of Rit Dye. The liquid dye seems to work the best and there is a full range of colors possible. Rit Dye is not food safe; therefore, the eggs dyed this way should not be eaten. You can experiment with the amount of dye to use, but the mix is generally half a teaspoon to two teaspoons of Rit Dye to one cup of water. Add two teaspoons of white vinegar to each color cup.

For a complete breakdown of colors and the mix ratios for Rit Dye, visit the website www.ritstudio.com/color-library/color-archive. This is a great alternative to the basic colors. Have fun experimenting by mixing colors and using textures and appliqués.

EGG IN A BOTTLE

Here is a fun little experiment for kids to amaze their friends.

Needed:
- 1 hard-boiled egg, peeled
- Pan of hot water (with adult's help)
- Bowl of ice water
- Bottle with mouth opening slightly smaller than egg
- A bit of vegetable oil

Lightly oil the inside edge of mouth of bottle. Place the bottle in the hot water for approximately two minutes. Remove the bottle from the hot water and place the hard-boiled egg on the mouth, pointed side down. Now place the bottle into the ice water. The egg will slowly be sucked down into the bottle.

To remove the egg, tip the bottle upside down so that the egg rests inside the neck of the bottle. Blow into the bottle with it upside down and the egg will be pushed back out.

Appendix C: Additional Resources

Henderson's Handy Dandy Chicken Chart
www.sagehenfarmlodi.com/chooks/

APA-ABA Youth Poultry Club
www.apa-abayouthpoultryclub.org

American Bantam Association
www.bantamclub.com/aba/

American Poultry Association
www.amerpoultryassn.com

Omlet - Chicken Breeds
www.omlet.us/breeds/chickens/

Veterinary Drug Database
www.drugs.com/vet/

Poultry Hub
www.poultryhub.org

BackYard Chickens Forum
www.backyardchickens.com

NetVet - The Electronic Zoo
http://netvet.wustl.edu/birds.htm

Poo - The Chicken Keeper's Guide
http://chat.allotment-garden.org/index.php?topic=17568.0

About the Author

With over forty years of backyard poultry experience, Eric Lofgren has a passion for chickens. This—combined with his broad knowledge and approachable writing style—has helped him create an informative and easy-to-follow guide for raising and caring for your backyard flock.

With extensive studies of backyard and commercial poultry-keeping practices from the United States and around the world, Lofgren has compiled his knowledge into words that even the younger poultry raisers can understand and easily follow.

Eric Lofgren currently resides in Central Florida, after spending the first thirty-seven years of his life in the Pacific Northwest.

Thanks

A special thanks to my mother, who, with her love and support throughout the years, has made this book possible.

Dedication

This book is dedicated to all the young boys and girls out there who have a dream of raising chickens. Don't give up on your dream.

Index

abbreviations, 15–17
aflatoxicosis, 187
airsacculitis, 216, 217
alektorophobia, 17
Ameraucana, 44–45
American Humane Association, 20
American Poultry Association (APA), 14, 47, 76, 160
anatomy, 21
    of an egg, 102
    feathers, 144, 145
    internal, 165, 192
    reproductive system, 99
    skeletal, 164
    spurs, 150
Animal Welfare Approved, 20
anti-pick sprays, 212
antibiotic free classification, 18
antibiotics, 214, 215, 216, 217
antiparasitics, 169, 171–173, 175, 217, 218
artificial lighting, 55–59
auto-sexing, 158–160
avian influenza, 188–189

bacterial infections, 167, 190, 197

Barred Plymouth Rock, 42–43, 157, 159
bathing, 27–28, 139–142
beaks
    clipping, 211, 212
    filing, 147, 148
    fixing cracks, 147, 148
    trimming, 146–148
Black-Sex-Link, 157, 158
blindness, 199, 203, 204
blood stop, 140, 141, 142, 150, 179, 191
boredom, 211
botulism, 205–206
breeds, 13, 14. *see also under breed name; genetics*
    auto-sexed, 158–160
    crossbreeds, 160–162
    heritage, 156, 157
    hybrids, 14, 157, 158
    mutts, 160, 161
    ornamental, 153
    overbreeding, 42, 156, 161, 183
    purebreds, 14, 158
    selection of, 39, 41, 47, 74
    sex-linked, 158–160
bronchitis, 197, 213
brooders
    ammonia levels, 80
    heat sources, 28

    litter, 78, 79, 80, 181
    moving chicks into, 113
    set-up, 28–29, 76–78
    space requirements, 77
    temperature, 62, 76–77, 81, 115, 121
    transferring chicks to pens, 30, 61–62, 82
broody hens
    breaking, 93–94
    moving, 94–95
    recognizing, 89
    setting, 13, 56, 90–91
    sitting, 13, 90
    surrogacy, 92–93
    time spent with chicks, 92, 93
Brown Leghorn, 159
bumblefoot, 190–193

cage free classification, 17
calcar bones, 149, 150
cannibalism, 210–212
certifications, 19, 20, 75
chicks, 13. *see also incubation*
    0 to 6 weeks, 79–82, 113, 115
    6 to 14 weeks, 82–84
    14 to 22 weeks, 84–86
    fall hatch, 119–122

feathering out, 81–82, 121
growth rates, 156, 157
handling, 79–80, 84
purchasing, 73, 75
transitioning to flock, 130–132
cholera, 214, 215, 216
chronic respiratory disease
   disease, 193–194
   treatments for, 194, 214, 215, 216, 217
classifications, 17–20
climate. *see weather issues*
coccidiosis, 80, 194–195, 215, 216
coccidiostat, 215, 216
cockerels, 13, 123
cod liver oil, 215
cold exposure, 60–64
combs, 40
composting, 70–72
conjunctivitis, 203
coops. *see also nest boxes*
   amenities, 26
   building materials, 50, 51
   cleaning, 49, 55, 167, 189, 197
   latches, 49
   roovess, 49, 50
   size, 26, 48

ventilation, 26, 49
Cornish Crosses, 46–47
coryza, 197–198, 214, 216
CRD. *see chronic respiratory disease*
crop flushing, 208–209
crossbreeding, 156, 160–162
curled toes, 176–178

debeaking, 19, 146, 147
Delawares, 43–44, 158
diseases, 163, 187. *see also under disease type*
   carriers of, 127, 193, 194, 197, 198, 200
   in chicks, 80, 194
   culling sick birds, 104, 189, 198, 212
   immunity to, 128, 194, 196, 200, 202
   stress related, 193
   testing program, 218–222
   transmission, 164–165, 167, 195
dry pox, 196
dust bath boxes, 27–28
dust bathing, 139, 171

E. Coli, 104, 167, 190
ear mites, 217

egg production, 56, 57, 96, 99
   collection frequency, 103, 104
   factors influencing, 153–155
   first laying, 87
   time from hatch to laying, 120
eggs. *see egg production; incubation*
   color versus taste, 100–103
   fertile versus nonfertile, 101, 108
   formation, 96–100
   freshness, 100, 101, 103–105
   receiving shipments of, 117
   selling, 22, 23, 24
   storage, 103–105
   USDA classifications, 17–20
   washing, 103, 104
ethylene glycol poisoning, 206–207
Exotic Newcastle Disease (END), 200–201

feather picking, 77, 211, 212
feathers, 144, 145

feed. *see also nutrition*
    amount of, 27, 34–35
    commercially produced, 183
    costs per bird, 47
    egg production, 155
    during molting, 152
    protein levels in, 35–37, 63, 65, 183
    starter, 78, 195
    storage, 188
    toxins in, 188
    types of, 27, 30–34, 63
feeders, 27, 29
fenbendazole, 169
fleas, 174–176, 217
flip-over syndrome, 184–186
flocks
    bringing in new birds, 126–130
    changing dynamics of, 132–133
    hierarchies within, 123–126, 136
    mixed-ages, 85
    territorial behavior, 90–91, 123, 125
    transitioning new birds, 130–132
Food Alliance certification, 20
food poisoning, 188
fowl cholera, 214, 215, 216
fowl pox, 195–196, 213

free range classification, 18
fungal infections, 201–203

gardening, 68–70
genetics, 155–156. *see also breeds*
    ailments, 176, 187
    color variations, 156, 157, 160
    and commercial use, 156
    and cross breeding, 161
    line breeding, 156
    trait selection, 156, 157–158
Gold Legbar, 159
government inspections, 218, 221
grooming, 139–142
growth hormones, 18

hatcheries, 75, 76, 119
heart problems, 185, 186–187
heat sources, 28, 62–63, 77
heat stress, 59, 65, 218
hens, 13. *see also broody hens*
    reaching maturity, 86, 87–88
    signs of readiness to lay, 146
herpes virus, 195, 199
H5N1, 188

hormone free classification, 18
humane certifications, 19
hybrids, 14, 157, 158

illnesses. *see ailments; diseases*
impacted crop, 207–209
incubation, 103, 105
    basics, 106
    candling, 110–112
    dos and don'ts, 115–117
    egg turning, 107, 109
    egg viability, 107–108, 117
    embryo development, 110, 111
    hatch rates, 106, 110, 119
    hatching, 112–113, 118, 119
    heat source, 108–109
    humidity, 109–111, 112
    lockdown, 112
    problems, 117–119
    temperatures, 107, 108–109, 115, 116
Infectious bronchitis, 196
Infectious Coryza, 197–198, 214, 216
infectious synovitis, 215, 217
influenza, 188–189

Leghorns, 45
legs
- bow-legged, 180
- hobbling, 181–182
- splints for, 180, 181

lice, 170–172, 218
life stages. *see also chicks; young adults*
lighting. *see artificial lighting*
lymphoid leukosis, 198, 199–200

Marek's Disease, 198–200, 213
meat birds, 14, 23–24, 31, 47
medications, 217–218
- capsules, 214–215
- contamination of eggs/meat, 213
- dosages, 212
- injectables and ingestibles, 213–214
- off-label use, 166
- reactions to, 213
- solubles, 215–217
- tablets, 214–215
- vaccines, 213
- withdrawal times, 166

mineral deficiencies, 182–184, 218
misters, 65
mites, 170–172, 218
mold, 201, 203, 205
molt, 15, 56
- and age of bird, 151
- and diet, 57
- forced, 18, 19, 56
- in roosters, 152
- timing of, 151, 152
- variables affecting, 151

mycoplasma gallisepticum, 193

nails, 145–146
nest boxes, 53
- height, 55
- keeping clean, 104
- location, 52, 55
- materials for, 27, 51, 52, 104
- size, 54

New Hampshire Reds, 42, 158
Newcastle disease, 200–201, 213
nutrition. *see also feed*
- deficiencies, 182–184
- low sodium diet, 187
- meat, 71, 211–212
- treats, 37–39, 82–83, 155

organic certification, 19
Oxine, 192, 196, 202–203

paralysis, 199, 205
parasites, 168, 217, 218
pasty butt, 80, 178–180
pens, 50
- design, 26
- disinfecting, 200, 203
- floorings, 71
- keeping out predators, 66 67
- temporary cages, 68, 130

personalities, 133–136
pesticides, 19, 171
poisoning, 206–207
poultry combs, 40
poultry labeling, 17–20
pox, 195–196, 213
Poxine-Fowl Pox, 213
predators, 26, 66–68, 153
pulmonary hypertension syndrome, 186–187

quarantines, 127, 129, 201

range paralysis, 199, 213
Red Cell, 218
red pepper flakes, 169
Red Star, 157
reproductive system, 96, 99
respiratory infections, 201–203. *see also chronic respiratory disease; coryza*
Rhode Island Reds, 41–42, 157, 158
riboflavin deficiency, 178
roosters, 13, 56
- aggressive, 126, 136–139
- establishing dominance over, 126, 137–139

fertility, 107
molting, 152
normal behavior, 136
spurs, 148–151
roosts, 48, 71
for chicks, 83
height of, 48, 190

Salmonella, 104, 167
scaly leg mites, 172, 218
selling birds, 22, 127, 162, 194, 198
Sevin dust, 171, 172, 175
sex-links, 14, 46, 158–160
show birds, 139, 140, 142, 210
sinusitis, 203–204
slipped Achilles tendon, 180
space requirements, 25, 74
splints, 181
spraddle leg, 180–182
spurs, 148–151
Staphylococcus, 167
stress, 127, 128, 153, 193
Sulfadimethoxine Soluble, 216
Sulmet, 195, 216

Sumatra roosters, 148
surrogacy, 92–93
synovitis, 215, 217

tapeworms, 213, 214, 215.
toes
　cleaning, 141
　curled, 176–178
　trimming nails, 145–146
transporting, 127, 128, 129, 142, 173

USDA classifications, 17–20

vaccines, 200, 213
vegetarian fed classification, 19
vitamin deficiencies, 178, 182–184, 218
vitamin supplements, 214, 215, 218
vocalizations, 125, 131, 132, 135

water
　for drinking, 27, 78
　in hot weather, 27, 64, 65
water belly, 186–187
waterers, 27, 29, 78, 164
weather issues, 25, 59, 155
　acclimation, 60–62
　cold, 60–64
　and drafts, 49, 62, 116
　heat, 64–65
　lethal body temperatures, 60
　sun exposure, 64, 210
　temperature
wet conditions, 194, 201–202, 203
wet pox, 196
White Rock, 157
wormers, 213, 214, 215, 216, 217, 218
worms, 71, 167, 169–170, 212
wound infections, 214, 215

young adults, 86–88

The Backyard Chicken Bible Copyright © 2014 by Eric Lofgren. Manufactured in the United States. All rights reserved. No part of this book may be reproduced in any form or by any electronic or mechanical means including information storage and retrieval systems without permission in writing from the publisher, except by a reviewer who may quote brief passages in a review. The content of this book has been thoroughly reviewed for accuracy. However, the author and publisher disclaim any liability for any damages, losses or injuries that may result from the use or misuse of any product or information presented herein. It is the purchaser's responsibility to read and follow all instructions and warnings on all product labels. Published by Living Ready Books, an imprint of F+W Media, Inc., 700 East State St., Iola, WI 54990. (800) 289-0963. First Edition.

Other fine Living Ready books are available from your local bookstore and online suppliers. Visit our website, www.livingreadyonline.com. Living Ready® is a registered trademark of F+W Media.

18 17 16 15 14   5 4 3 2 1

ISBN-13: 978-1-4403-3924-0

Distributed in Canada by Fraser Direct
100 Armstrong Avenue, Georgetown, Ontario, Canada L7G 5S4, Tel: (905) 877-4411

Distributed in the U.K. and Europe by F&W Media International, LTD
Brunel House, Forde Close, Newton Abbot, TQ12 4PU, UK,
Tel: (+44) 1626 323200, Fax: (+44) 1626 323319, E-mail: enquiries@fwmedia.com

Distributed in Australia by Capricorn Link
P.O. Box 704, S. Windsor NSW, 2756 Australia, Tel: (02) 4560-1600,
Fax: (02) 4577-5288, E-mail: books@capricornlink.com.au

Edited by Amy Owen and Kelsea Daulton
Cover design by Christy Cotterman
Interior design by Laura Spencer
Illustrations by Patrick Welsh
Production coordinated by Debbie Thomas

FREE COMPOSTING TIPS FOR HAPPY CHICKENS

Did you know you could put your chickens to work? It's true—one of those ways is by letting them tend to your compost heap. Check out these free composting tips at http://www.livingreadyonline.com/free-download-composting-tips/

MORE BOOKS ON SELF-SUFFICIENCY

Build The Perfect Bug Out Bag
By Creek Stewart

Living Ready Pocket Manual: First Aid by James Hubbard, The Survival Doctor™

Recipes & Tips for Sustainable Living by Stacy Harris

AVAILABLE ONLINE AND IN BOOKSTORES EVERYWHERE!

To get started join our mailing list at www.livingreadyonline.com.

Become a fan of our Facebook page: facebook.com/LivingReady